FROM THE GULF

TO

ARARAT

G.E. HUBBARD

FROM THE GULF

TO

ARARAT

IMPERIAL BOUNDARY MAKING IN THE LATE OTTOMAN EMPIRE

FOREWORD BY

RICHARD SCHOFIELD

New edition published in 2016 by
I.B.Tauris & Co Ltd
London • New York
www.ibtauris.com

First published by William Blackwood and Sons in 1916 as *From the Gulf to Ararat: An Expedition through Mesopotamia and Kurdistan*

ISBN: 978 1 78453 121 8

A full CIP record for this book is available from the British Library
A full CIP record is available from the Library of Congress

Library of Congress Catalog Card Number: available

Printed and bound by CPI Group (UK) Ltd, Croydon, CR0 4YY

For Bob who, like his grandfather, never lowered his sights in the face of adversity

CONTENTS

ILLUSTRATIONS

Introduction by Susan Littledale

Plate Section

The following captions are taken word-for-word from G.E. Hubbard's original photo albums.

Endpapers

Jacket images

FOREWORD

By Richard Schofield

With the current centenary of World War I, and a depressing intensification of human suffering in many conflict-ridden regions of the Middle East, questions surrounding the legitimacy of the framework of state territory laid down by Britain and France at the turn of the 1920s have only heightened. The collapse of the sovereign reach of the state – first in Iraq, then more recently in Syria, Libya and Yemen – and the emergence of new territorial structures and power spatialities in the state margins have led some to observe the obsolescence of these colonial borders. In purely spatial terms, the emergence of Islamic State is intriguing, for what we have witnessed in the last couple of years is the deliberate territorial consolidation of a trans-boundary entity joining the wider borderland areas of Syria and Iraq – a would-be trans-boundary borderland state if you like. In an era where many international boundaries have been securitised (and, indeed, fortified) by states against unwanted human movement, interest in Middle Eastern borderlands as the 'spaces in between'[1] the territorial reach of states has extended to a far wider audience of policy makers and academics than is customary. Yet in all of this, at least in broad terms, the security community is equating conflict borderlands with those spaces (usually marginal) that are beyond the control of state governments. Whatever the terminological niceties, what these heightened degrees of scrutiny have perhaps highlighted is that we don't always know as much about regional borderlands as we sometimes assume – and that goes particularly for their materialities and histories: how states have attempted to extend and project their power and how this has been resisted locally.[2]

In truth, there aren't that many historical Middle East borderlands in the classical sense of the term – that is zonal states or imperial margins that have served as a frontier. The lines laid down as borders by Britain and France in the aftermath of World War I were by no means purely arbitrary and many reflected vague and putative former Ottoman divisions, but they had little to no tradition or function as political boundaries, frontier zones or border marches. Except, that is, for the Ottoman-Persian border margins which, by any reckoning, constituted classic historical borderlands. Although overstated in obviously deterministic terms, there is something to be said for our author's characterisation here:

> sooner or later this great mountain range [the Zagros] has invariably resumed its original role as a divider of nations.[3]

Although this boundary was supposedly settled and laid down on the ground during 1914, as attested to in this new edition of G. Ernest Hubbard's wonderfully readable border travelogue, this was the long overdue culmination of a tortuous (joint Anglo-Russian) imperial project to narrow a traditional frontier. So, while separated by only half a decade, the 1914 Ottoman-Persian border demarcation, and the introduction of the 'Sykes-Picot system' of borders to define Britain and France's new protégé states at the turn of the 1920s, were worlds apart in conception and temporal calculation.[4]

This new I.B.Tauris edition of Hubbard's classic work benefits most obviously from the illuminating introduction penned by his granddaughter Susan (Sue) Littledale. Those, like me, who have cherished *From the Gulf to Ararat* for a good while, now know much more of the man and the personal context behind the book. We are also lucky to have had access to the family archive via Sue, which has allowed the inclusion of a good number of fascinating photographs beyond those included in the original 1916 publication. More on that soon but the launch of this new edition delights me for both personal and professional reasons. When researching my Master's dissertation at the University of Durham during the late spring of 1984,[5] I bumped into my fellow Lancastrian (and future SOAS colleague) R. Michael Burrell on the newly carpeted steps of the SOAS library.[6] On mentioning that I was trying to piece together a historical geography of the Shatt dispute – nominally, at least, a stated cause of the then ongoing Iran-Iraq War – Mike came back with the following firm retort in his inimitable Blackburn accent:

> There's one historical source you must read above all others – *From the Gulf to Ararat* by a certain Mr G.E. Hubbard. Not only is it good and seriously underestimated but it's beautifully written and riotous fun.

I continued up the steps and made my first acquaintance with the work twenty minutes later.

Mike sadly passed away too early in the late 1990s. In addition to the personal qualities I have endnoted here, he was known for his incredible private book collection and, in his latter years, for eloquently capturing a moment with a few well chosen words to bring sessions of RUSI's Middle East Forum to an entertaining close. It had been an article of his, co-written with yet another charismatic North-Westerner, Keith McLachlan, which had originally got me hooked on the Iran-Iraq borderland as a geography undergraduate at SOAS.[7] With that irrepressible glint in his eyes, Keith inspired many students to research the Middle East during his long, distinguished career in SOAS Geography, with a passion that was never more evident than when he talked about his beloved Iran. He lectured and wrote about Iran's territorial issues old and new with real insight. As time progressed, I was lucky enough to collaborate with Keith in researching such matters and if I've had some success in covering regional boundary questions over the years and decades, much of the credit goes to him as a wonderful mentor. Keith died in October 2013 and, like Mike, is sorely missed. In a small but entirely suitable way, I like to see this new centenary edition of *From the Gulf to Ararat* as a tribute to both of them and their inspiring legacy. They'll be grinning up on high!

One of the things that drew my attention on first flicking through the pages of Hubbard's work that afternoon back in 1984 in the SOAS library was the photograph of the British delegation to the 1914 Turco-Persian Frontier Commission (with their dogs!), opposite page 16 of the original edition. In their civvies, Hubbard (GE) and a young Arnold Wilson obviously stood out from their contemporaries but they also looked somehow more worldly and intent, if slightly elusive. Of course, Wilson would go on to be (with Gertrude Bell) Iraq's chief imperial architect, and Hubbard the resourceful chief technician who effectively facilitated the 1914 Commission. But those first impressions of elusiveness have stuck with me. Chatting with Sue and reading her introduction have, of course, helped me form a much rounder picture of our author but my feeling persists that there was still something of a 'man of mystery' about GE.

Anyway, it was soon apparent to me that *From the Gulf to Ararat* was everything Mike Burrell had said about it – it was pacy, pithy, observant, humorous, level-headed and, in some ways, refreshingly down to earth. But what of its academic merit as a chronicle of boundary-making and as a study of borderlands in the early twentieth century? Here it sits well with the better-known, British-penned geographical works of the era from Lord Curzon, L. W. Lyde, Thomas Holdich and Charles Fawcett. First, it was the story of how Britain and Russia finally drew a line under the protracted saga of narrowing a traditional frontier, a venture they had prioritised some seven decades earlier. Second, it was penned by an individual who was largely responsible for ensuring that the job was done. While GE was merely Secretary to the British Delegation and Lieutenant-Colonel C. H. D. Ryder its Chief Surveyor, it was these two individuals who were largely credited with the achievement of getting a line down on the ground during 1914.[8] Arnold (then Captain) Wilson, Britain's Deputy Commissioner, suggested at the end of the year that if the final decisions on delimitation and mapping had been Ryder's, then the fact that the whole exercise was completed in double-quick time before World War I made itself felt in the region was more down to GE, who effectively served as secretary to the whole commission and not just to the British delegation.[9] So the finer details and niceties of the 1914 line that survives to constitute the legal basis of Iran's contemporary boundary delimitations in the west with Turkey and Iraq could be attributed to Ryder's handiwork and GE's facilitation.

In only the second chapter of *From The Gulf to Ararat*, Hubbard soon displays that he was as conversant as any of his fellow British boundary-drawers with the latest (Curzonian, if you like) thinking in the top-down science of imperial boundary-drawing and its preference for the adoption of natural features:

> Of physical features there is hardly one commonly used in frontier-making which we did not, at some time or other, apply: along the Shatt-el-Arab, 'the line of low tide'; in the case of other rivers, one or other bank, or oftener still the '*medium filum aquae*'; in mountainous regions 'the crest line' or 'the watershed' (by no means synonymous terms), or else that much disputed phrase 'the foot of the hills'.[10]

Yet any such preferences were qualified – for the realities of quickly laying a line down on the ground along such a long (all 1,180 miles worth) varied and complex borderlands necessitated a pragmatic flexibility – and 'horses for courses' seem to have been the governing watchwords here.

> There is one noticeable characteristic of the Turco-Persian frontier which is due partially to the piecemeal and deliberate way in which it has been created. It supplies instances of practically every principle of delimitation known to the science. A frontier may be geographical, racial, linguistic, religious or even purely artificial. That in question is all of these. In the broadest sense it is geographical, for it follows in the main a great mountain range.[11]

And so GE continued: the overriding basis of the 1914 delimitation was racial in the south – separating Lur from Arab; linguistic in the Pusht i-Kuh region of the central borderlands; and alternately tribal and linguistic further north in Kurdistan. It was admitted that the line in the alluvial flats of Khuzestan that join the Shatt al-Arab in the far south was 'frankly artificial', 'fixed according to the "astronomical method" adopted so generally in Africa and North America'.[12] We also know that strategic calculations and expediency also dictated occasional departures from the overall basis of delimitation agreed in the November 1913 Constantinople Protocol. GE alludes here to the sovereign enclaving of oil wells in the central stretches of the borderlands and the deliberate balancing in possession of strategic mountain passes further north.[13] Acknowledgement of such complexity would also make something of a mockery of the many rather simplistic boundary-type generalisations and classifications that would follow in academic boundary studies in the interwar years, especially those produced by those famous American geographers, Richard Hartshorne and Samuel Boggs.[14]

Hubbard reminds us that individual outcomes in the imperial boundary-drawing process varied considerably in practice, largely depending on the proclivities of the individuals charged with such tasks but also on the resources thrown at the question at hand. There are other ways in which he guards against any cosy assumptions about the colonial boundary stereotype. Though his book revolves around the central act of laying down a line in stone, it tells us much about the surrounding borderlands in zonal terms, with important observations made about the human groups that inhabit them. Frequently perceptive coverage of the local socio-economic borderscape neatly counterbalances the official character of the imperial mission, and the encounter between top-down officialdom and ground-up local borderlander, recalled vividly and entertainingly by GE, is one of the most fascinating aspects of the work. On page 2 Hubbard introduces the summary introductory chapter to come, with its aim to provide

> an example (though happily not a very typical one) of the way a frontier comes into being.[15]

Sometimes, the book demonstrably exceeds this modest ambition – showcasing how a border actually operates. This is true of the new edition in its provision of supporting photographs beyond those included in the original publication. There is a wonderful sequence of images entitled 'The Lost Caravan' that showcase varying degrees of order and disorder within human traffic at the border – GE pencils captions to accompany the final three photographs of the sequence 'Confidence', 'Doubt' and 'Confusion', as near chaos quickly and progressively descends (plates 34–6). Here I am immediately reminded of the striking visualities present along many contemporary borderscapes, particularly backlogs of traffic at the more securitised boundaries, such as those separating Israel and the West Bank, or the United States from its neighbours to the north and south.

Also worthy of mention here is the catalogue of continuing misfortune suffered by the British delegations to both the mid-nineteenth and early twentieth century commissions charged with setting the Ottoman-Persian boundary down in stone. GE himself suffered a serious shooting injury in the course of his duties as Secretary to the British Delegation late during proceedings in 1914, reflected upon at some length by the author and also his granddaughter Susan (Sue) Littledale in her introduction to this new edition. As mentioned, there had been a tradition of bad luck, most notably when the ship carrying the British Commissioner back from the 1849–52 quadripartite border survey got all the way back into internal British waters but then sank at Gravesend. The individual concerned, Colonel Fenwick Williams, survived to tell the tale but he had been the fastidious note-keeper on this mission and getting on for three-quarters of the materials that were going to be used for resultant borderland mapping, not to mention his own report of proceedings, had been lost to a very muddy River Thames.[16] Luckily, the detailed travelling diary maintained by Williams' Russian counterpart, Colonel Y.I. Tchirikof, survived but the mapping effort had been seriously jeopardised from the start by such a catastrophic data loss.

However well GE's book hangs together by itself, we need to reiterate that it entertainingly recalls the conclusion to a long, confused and tortuous imperial venture by which a frontier was ostensibly narrowed to a line – in GE's own, memorable words, 'a process of procrastination unparalleled even in the chronicles of Oriental diplomacy'.[17] Hubbard sheds light on the prevailing socio-economic geography of the wider borderlands, with important commentary passed on the state of Baghdad and Basra a good half-decade before the inception of Britain's mandated state territory of Iraq in 1920. Neither do the unfolding horrors of the Armenian genocide escape his attention. Yet I still hold that the book's central concern – the 1914 Ottoman-Persian boundary demarcation – was in essence a mid-nineteenth century exercise in conception and execution, which just so happened to materialise 70 years later than it should have done. Let's now provide some context for such a statement.

As I commented a few years back, the Perso-Ottoman borderlands had for centuries constituted a classic frontier zone:

> Classic in the sense that they existed as a fluctuating, conflict-prone zone to which competing imperial authority was never extended on a permanent basis; also classic in the sense that for the Ottomans they constituted remote and unfamiliar terrain and for the Persians, the divide separating the aliens to the west. Finally this textbook traditional frontier provided ample opportunities for localised political autonomy and the relative freedom of socio-economic borderlands movement.[18]

A rare half-decade of cooperation between Europe's big 5 moving into the 1840s had persuaded Britain and Russia that the time was right to stabilise a traditional frontier that now mattered to them. Russia had only recently expanded its borders into the southern Caucasus at the expense of Persia, while Britain wanted to push its developing economic interests at the opposite end of the borderlands more determinedly into Mesopotamia. Both therefore saw advantages in narrowing a traditionally insecure frontier into a mappable (and hopefully more permanent) line. Yet freezing a moment in time, or the territorial balance it estimated as prevailing along the borderlands in 1843, Britain's rationale for stabilisation was not only flawed in conception but proved impossible to establish on the ground. While the breakthrough May 1847 Erzurum Treaty would introduce a delimitation of sorts (more the bones of a territorial deal in truth) from the Shatt al-Arab in the south to Mount Ararat in the north, the resultant efforts of Britain, Russia, the Ottoman Empire and Persia to lay down a line on the ground would consequently founder on such misplaced, grandiose ambition. If the European powers would soon demonstrate that their knowledge of a complex borderland was inevitably superficial, then the regional imperial powers didn't always display a great familiarity with its dynamics themselves. It is fair to say that the prioritised European project to narrow the frontier was, where practicable, utilised by the sovereign protagonists to their own perceived benefit. First, the Ottoman Empire saw the venture as a vehicle for materially improving its territorial definition in the remote east to the detriment of Persia. Somewhat defeatist if notably pragmatic, a relatively weaker Persia believed it would do better by participating in the European project than by confronting Constantinople directly.[19]

Not possessing the mandate it required to railroad through a territorial settlement also predictably dictated that no Perso-Ottoman boundary was laid down on the ground during the nineteenth century. For Britain and Russia were mediating powers only. The plans of the quadripartite 1849–52 Turco-Persian Boundary Delimitation Commission to firm up the 1847 Erzurum delimitation after ground surveys, and to demarcate that treaty line, were soon relegated to mapping a frontier zone in which the boundary lay. Both the Ottoman and Persian Commissioners maintained widely divergent interpretations of the 1847 line and Britain and Russia possessed neither the authority nor the coercive skills to reconcile them to the line they had meant to introduce. By 1869 a lavish Anglo-Russian Identic Map (*Carte identique*) of the borderlands measuring some 60 feet in length (and drawn at a scale of 1 inch to 1 mile) had been produced on the basis of the 1849–52 survey, depicting a frontier zone generally varying in width from 20 to 50 miles.[20] Some 4,000 discrepancies had to be ironed out before Britain and Russia finally seemed set to get a solid indication of where the Ottoman and Persian governments saw the boundary as lying within such a zone. Reputedly, a gold Ottoman line was traced in its eastern reaches and a Persian silver equivalent in its westernmost margins, effectively defining the Identic Map zone as an area of overlapping sovereign claims. Before Britain and Russia could get any further in following things up, the Serbian War broke out in 1877 and that was the end of that. No boundary would be agreed, though the unavailing efforts of the European imperial powers to narrow the frontier had left their mark in another way:

> one consequence of this [...] project was to progressively instil a more developed territorial consciousness or defensiveness in the local disputants, whereby territorial definition in itself would become a major component of identity and rivalry.[21]

Moving into the twentieth century, the discovery of oil in south-western Iran only deepened Britain's fast-developing economic and strategic interests in the northern Gulf region and along the Shatt al-Arab.[22] Demonstrably weaker than a half-century previously, the Ottoman Empire was now prepared to be much more accommodating and pragmatic than in its earlier dealings over the boundary question.[23] Still, it secured a broadly favourable territorial settlement with the November 1913 Constantinople Protocol, pretty much retaining its sovereign rights over Shatt waters. Crucially, however, it was the significant empowerment of Britain and Russia that enabled the 1914

Turco-Persian Frontier Commission to complete its task in 10 months flat during the year:

> Such progress could be attributed to three factors: its possession of decisive arbitrary powers, unusually favourable weather conditions; and the time spent beforehand thrashing out an unambiguous territorial deal.[24]

The ability to thrash out a more detailed and viable territorial deal reflected the reality that more was known about the way in which the borderland and its inhabitants operated after all of the consideration of the previous six decades. At least to a degree, all sides had learned from their previous mistakes, meaning that the territorial deals comprising the first stage of delimitation in the 1913 treaty settlement were much better informed and workable than those specified for the borderlands in the 1847 deal. Perhaps Britain was also getting slightly better at drawing boundaries – after all, it had gained plenty of experience in the intervening period!

> By this stage, too, both of the local powers were fully consonant with Westphalian precepts of sovereignty and had done much to territorialise their margins, making acceptance of a largely European territorial blueprint much the easier.[25]

Maybe a final consideration paving the way for the 1914 Commission's speedy progress was that sheer weariness had set in on all sides, fostering a determination to finally draw a line under the whole business.

The 1914 Turco-Persian Frontier Commission recorded its detailed description of the line it laid down in its Proceedings (*Procès-verbaux*), as detailed in the next paragraph. GE had been very much in charge of their every aspect. Tracings of their finalised delimitation were then made on 10 (reduced) sets of the 1869 Anglo-Russian *Identic Map (Carte identique)*.

> The commissioners also plotted their delimitation upon ten new sets of a *Carte Supplementaire*, prepared on the basis of surveys they undertook themselves to illustrate the border zone in sections that were shown with neither sufficient detail nor accuracy by the *Carte Identique*.[26]

To gain a more complete picture of how the Ottoman-Persian boundary was settled in 1914, GE's book should ideally be read in conjunction with the formal record of the 1914 Commission, the 'Proceedings of the Turco-Persian Frontier Commission, 1913–14', which he had essentially prepared and drafted. These were originally printed in French at the end of the 1914 demarcation effort, as 'Recueil des Procès-Verbaux des Seances de la Commission de Delimitation de la Frontiere Turco-Persane, 1913–1914' in a detailed 194 page document.[27] A translated summary was later produced by Britain's Office of the Civil Commissioner in Mesopotamia in 1920.[28] A more digestible account is provided in Arnold Wilson's detailed but readable report of the workings of the Commission for the period (from mid-July 1914 onwards) he acted as Commissioner because of the enforced absence of Albert Wratislaw.[29] The later musings of British surveyor Colonel Ryder on the 1914 exercise are well known as a consequence of his much referred to 1925 piece in the *Geographical Journal* but it is also instructive to read his original short report, also produced at the year end of 1914.[30]

Of course, this exclusive reliance upon the official British record might reasonably be questioned. While this is understandable in as much as the evolution of the boundary from the mid-nineteenth century through to World War I was largely Britain's show, there has been an encouraging trend of recent research into parallel Persian and Ottoman historiographies. The works of Firoozeh Kashani-Sabet, Sabri Ates and Burcu Kurt come most readily to mind here.[31] This is also consonant with the current increased levels of interest being shown in borderlands and their historical materialities within the humanities and the social sciences more generally, as alluded to at the beginning of this foreword.

I think there is no more appropriate way to end here than to reproduce, in full, Arnold Wilson's very full acknowledgement of the contribution GE made to the 1914 demarcation exercise.

> I am indebted to this officer throughout the Mission for willing, cordial and valuable assistance rendered to me first as Deputy Commissioner and later as Commissioner.
>
> Not only did he undertake a great deal of office work connected with the British Commission, but he was from the beginning practically solely responsible for the preparation, drafting and copying of all the proces-verbaux with their voluminous annexes. There was no Russian Secretary, the Persian Secretary was quite useless, seldom putting pen to paper and the drafts of the Turkish Secretary, besides being illegible, generally needed a good deal of tactful editing before they were fit to be placed on record.
>
> His tactfulness and energy, coupled with his command of French, won for him from the beginning a position in the Secretariat akin to that of Colonel Ryder in the 'corps technique'. By common consent he was given custody of the Central Archives and entrusted with the preparation of the fair copies (12 in all) of each process-verbal for distribution at meetings. He and Colonel Ryder were solely responsible for the detailed description of the frontier, which was accepted by all parties practically without alteration. But for the industry and enthusiasm displayed by him and by Colonel Ryder in their departments, the Commission would certainly not have finished this year: as it is, so long as he was with us on no occasion were we delayed for an hour for lack of documents: his departure was a real loss to the whole Commission, the foreign Commissioners only realising after his departure the extent of their dependence on him.[32]

This was a sincere and fulsome tribute and one which pinpointed our author's critical contribution as the chief facilitator of the 1914 demarcation mission.

Yet his portrayal and characterisation here, as with Ryder, is almost as a technician. Surveyors and secretaries would be the unsung heroes of many imperial boundary surveys and commissions. Few, if any of them, would write with the eloquence, humour and perception of G.E. Hubbard.

Richard Schofield
Department of Geography
King's College, London
2015

Notes

1 J. Goodhand (2008), 'War, peace and the places in between: why borderlands are central,' in M. Pugh et al. (eds), *Whose Peace? Critical Perspectives on the Political Economy of Peacebuilding*, Palgrave, London, pp. 225–44.

2 R. Schofield, C. Grundy-Warr and C. Schofield (forthcoming) *Contested Border Geographies: History, Ethnography and Law*, I.B. Tauris, London.

3 G. E. Hubbard (1916), *From the Gulf to Ararat*, William Blackwood and Sons, Edinburgh, p. 3.

4 I refer to the Sykes-Picot system to acknowledge that many important facets of the famous 1916 Anglo-French agreement to partition the land territory between the Eastern shores of the Mediterranean and the northern reaches of the Persian Gulf never came to fruition. Notably, Mosul province did not end up falling under a buffer territorial wedge of French control to separate Britain and Imperial Russia, while no international regime was arrived at for Jerusalem.

5 Later published as a book: R. Schofield (1986), *Evolution of the Shatt al-Arab Boundary Dispute*, MENAS Press, Wisbech, UK.

6 Mike taught Middle Eastern history there and was loved by his students for his generosity of time and spirit, encyclopaedic knowledge and, perhaps above all, those meticulously researched and carefully typed reading lists.

7 R. Michael Burrell and Keith McLachlan (1980), 'The political geography of the Persian Gulf', in A. Cottrell (ed.), *The Persian Gulf States: A General Survey*, Johns Hopkins University Press, Baltimore, p. 122.

8 R. Schofield (2008), 'Laying it down in stone: delimiting and demarcating Iraq's boundaries by mixed international commission', *Journal of Historical Geography*, 24, pp. 397–421.

9 'Report on the Proceedings of the Turco-Persian Frontier Commission from July 16 until its termination on October 26, by Captain A.T. Wilson, CMG, Political Department Government of India' in IOR file: *L/P&S/10/522*. Reproduced in Schofield (ed.) (1989), *op. cit.*, pp. 15–51. Though deputy to British Commissioner Albert Wratislaw, Wilson would serve as Commissioner from mid-July, after Wratislaw was forced to retire.

10 Hubbard (1916), *op. cit.*, p. 20.

11 Ibid., pp. 19–20.

12 Ibid., p. 20.

13 Ibid., pp. 20–1.

14 Yet their boundary-type classifications have, with time, proved to have a real utility in potentially highlighting what may have been in the boundary-drawer's mind. This is particularly valuable in instances where the postcolonial state is struggling with vague and ambiguous colonial boundary definitions.

15 Hubbard (1916), *op. cit.*, p. 2.

16 R. Schofield (2008), 'Narrowing the frontier: mid-nineteenth century efforts to delimit and map the Perso-Ottoman boundary', in R. Farmanfarmaian (ed.), *War and Peace in Qajar Persia*, Routledge, London, p. 160.

17 Hubbard (1916), *op. cit.*, p. 2.

18 Schofield (2008), 'Narrowing the frontier', p. 150.

19 For more detail on all of this, see Schofield (2008), 'Narrowing the frontier', pp. 149–73.

20 For wonderful quality images of sections of the 1869 *Identic Map*, see P. Barber and T. Harper (2010), *Magnificent Maps: Power, Propaganda and Art*, British Library, London, pp. 116–17.

21 Schofield (2008), 'Narrowing the frontier', p. 149.

22 R. Schofield (2004), 'Position, function and symbol: the Shatt al-Arab dispute in perspective', in L. Potter and G. Sick (eds), *Iran, Iraq and the Legacies of War*, Palgrave, New York, pp. 29–70.

23 Schofield (2008), 'Laying it down in stone', pp. 414–20.

24 Ibid., p. 417.

25 Ibid., p. 421.

26 Ibid., p. 418.

27 Available in the following India Office Record (IOR) file: *L/P&S/10/522*. Reproduced in R. Schofield (ed.) (1989), *The Iran-Iraq Border, 1840–1958*, Archive Editions, Farnham Common, vol. 6 (Demarcation of boundary by Mixed Commission of 1914 and border disputes following the Great War, 1914–1928), pp. 55–250.

28 See IOR file: *L/P&S/10/932*. Reproduced in Schofield (ed.) (1989), *op. cit.*, pp. 423–513.

29 'Report on the Proceedings of the Turco-Persian Frontier Commission from July 16 until its termination on October 26, by Captain A.T. Wilson, CMG, Political Department Government of India' in IOR file: *L/P&S/10/522*. Reproduced in Schofield (ed.) (1989), *op. cit.*, pp. 15–51. Wratislaw himself soon produced a useful note on the 1914 demarcation of the southern end of the borderlands late during December 1914: 'Note by Mr. Wratislaw on the demarcation of the Turco-Persian Frontier from Mohammerah to Vazneh'. See IOR file: *L/P&S/10/522*. Reproduced in Schofield (ed.) (1989), *op. cit.*, pp. 253–64.

30 'Turco-Persian Frontier Commission' by Lieutenant-Colonel CHD Ryder in IOR file: *L/P&S/10/522*. Reproduced in Schofield (ed.) (1989), *op. cit.*, pp. 266–76; C. Ryder (1925), 'The demarcation of the Turco-Persian boundary in 1913–14', *Geographical Journal*, 6, pp. 227–42.

31 F. Kashani-Sabet (1999), *Frontier Fictions: Shaping the Iranian Nation, 1804–1946*, Princeton University Press, Princeton and I.B.Tauris, London; Sabri Ates (2013), *The Ottoman-Iranian Borderlands: Making a Boundary, 1843–1914*, Cambridge University Press, Cambridge; Burcu Kurt (2014), 'Contesting foreign policy: Disagreement between the Ottoman Ministry of Foreign Affairs and the Ministry of War on the Shatt al-Arab dispute with Iran, 1912–13', *Iranian Studies*, 47, 6, pp. 967–85.

32 'Report on the Proceedings of the Turco-Persian Frontier Commission from July 16 until its termination on October 26, by Captain A.T. Wilson, CMG, Political Department Government of India' in IOR file: *L/P&S/10/522*. Reproduced in Schofield (ed.) (1989), *op. cit.*, pp. 15–51 (typed concluding section entitled 'Concluding remarks').

INTRODUCTION

By Susan Littledale

There is something magical about holding in your hand a letter that was penned in a distant and unfamiliar place more than a hundred years ago. That pleasure was enhanced for me when I first opened the old buff folder, tied up with ribbon, which contained a whole year's worth of such letters written by my maternal grandfather. Most of these missives, which spanned the months he spent travelling from the Gulf to Ararat, were written to his wife Dorothy, whom he had married a few short weeks before setting out on his journey.

The book you are about to read contains much that is taken from these letters and I leave my grandfather to tell the story of the expedition, but there is another story that he didn't tell. It is his personal story of the events that took him to its starting point and beyond: from his roving childhood as the second son of a peripatetic vicar, his early education and travels in Europe, and his place in an extended family exposed to the vagaries of life facing those who seek their fortunes in foreign climes.

On another level the wider collection of letters that my grandfather wrote home between 1904 and 1914 provides a unique archive of the experiences of a young recruit in the Levant Consular Service at a crucial time in the history of the Middle East. They begin when he was still an undergraduate at Trinity College, Cambridge. They end in August 1914 with his last letter from Bazirga in the mountains of northern Kurdistan.

With the help of the Hubbard family archive and a number of other primary and secondary sources, I have done my best to recapture both the man and the world in which he lived exactly one hundred years ago.

Susan Littledale
August 2014

Gilbert Ernest Hubbard, better known as G.E. Hubbard in his professional life, was born in Lower Beeding in Sussex on 27 April 1885. His father Cyril, a

Church of England vicar, had moved to the living in this small Sussex village two years earlier, shortly before the death of his own father, William Egerton Hubbard. The living was in the gift of William, a Russia merchant who had bought the nearby Leonardslee estate in 1853 after many years running the family business in St Petersburg.

The following year the family moved again, this time to Staffordshire where Cyril had been appointed vicar of St Michael's in Lichfield. GE remembered his father as a sincere and conscientious parson, although 'he always managed to get into quarrels with his congregation and his bishops and made life difficult for himself and my mother'. As was often the case with younger sons, Cyril had been marked down for the church at an early age 'although he ought never to have been anything of the sort'. GE, observing this sadness in his father's life, was determined from an early age not to fall into the same trap.

At the age of 7, GE was sent to boarding school in Sussex, where he remembers being 'quite happy' if 'a bit too much of a swot'. Holidays were spent in Lichfield, and later in East Sussex where the family moved when his father decided to give up parish work and become a small-scale farmer. The move was made possible because family mills in Russia, in which he had

1. G.E. Hubbard

capital, were for a short while paying out enough for him to indulge his love of countryside pursuits.

For GE this was also a time of significant change. After leaving his prep school at the age of 14 he had expected to follow his older brother, Leonard, to Charterhouse School. However, this was not to be: unexpectedly stricken with a childhood illness, he was instead to spend the next two years at home in the small village of Hartfield. He continued his education at a crammer in nearby Uckfield.

In retrospect GE saw this as a defining time in his life. He never did go to Charterhouse and, although he remembers the years in Hartfield as happy ones, they also marked him out as being different from his contemporaries – a difference explained both by the enforced loneliness of his teenage years and a natural seriousness of mind that belied his youth:

> I always felt rather a lonely sheep, not properly fitting into any of the ordinary cliques and groups of my contemporaries [. . .] but of ambition I had, I should say, more than the average amount. That is to say I was never content with a humdrum career, but had my eye pretty well fixed on the top of my particular tree.

When GE was just 16 the family fortunes took a plunge and his father was forced to look for another parish. By this time there were six children, the youngest just 1 year old. Leonard was at Cambridge but the other family members were uprooted from their comfortable way of life when Cyril took the position of chaplain in Gotha in Germany. GE's mother, Agnes, also set to work in Gotha, taking in English girls, who shared lessons with their own children. She was later to set up a school in France to provide an education for GE's three younger sisters and to augment the family coffers.

There was no place for GE in this arrangement and he couldn't be educated in Gotha. For the next two years he was sent to live with a German family in Dresden while receiving private coaching from an English tutor. He went home to Gotha in the holidays but it wasn't, as he later recalled, much of a life for a young boy of his age to be living with 'a crowd of rather neurotic females'. It's not surprising, therefore, that his imagination was soon gripped by a visiting family friend who had contacts with the Levant Consular Service and told of the opportunities it offered young men of high linguistic ability. It would, both he and the family agreed, be 'a good career' for him.

From this moment GE had the service in his sights and he spent much of the next seven years preparing himself to sit the competitive examination that was the only means of entry. Linguistic ability was the main criteria for applicants, who were expected to offer at least three, but preferably four, modern languages at a high standard in order to be accepted.

Thus it was that, after further sojourns in Rennes and Paris to brush up his French, GE set out on the career path that was to lead to his appointment as Secretary to the Boundary Commission ten years later. In 1903 he went up to Cambridge to study classics but this was just a necessary building-block in the pursuit of his final goal. He continued his travels around Europe in his vacations. In the summer of 1904 he 'engaged a nice curate' in the city of Angers to give him lessons in French. Subsequent vacations were spent in Florence where 'I am settled with a family [...] and have found a first-rate man to teach me' and many hours were passed in an exhaustive round of art galleries, museums and churches. He also made further visits to other European cities, reminiscent of the 'grand tour' of earlier generations albeit on a somewhat less generous budget. For GE, however, they weren't leisure pursuits: they were all part of the wider plan.

After graduating in 1906 he spent two more years abroad preparing himself for the open competition for the consular service. He sat the exams in 1908 and came fifth. There were just five places available. He was appointed Student Interpreter in the Levant on 30 May 1908. The palpable relief at discovering that he had made it after all those years of study can only be imagined.

Once accepted, GE completed two more years of intensive study, this time in Arabic, Persian and Turkish as well as history and law. It was not until 1910, after a short induction period in the Constantinople mission, that he was finally given his first posting as Acting Vice-Consul in the Bulgarian capital, Sofia.

* * *

There was a plague in Turkey when GE left Constantinople in mid-October. At the frontier he and his fellow travellers were 'bundled out into the bare countryside' and 'herded into quarantine sheds' where they had to stay for a couple of days before being allowed into the country. Once back on the train, with the carriages locked and guarded, he travelled through to Belgrade, the Serbian capital, before taking another train back to Sofia.

Tensions in the region were high. Bulgaria had recently achieved full independence from the Ottoman Empire although large elements of its ethnic population remained under Ottoman rule in Macedonia. Indeed, there was resentment of Turkey's continued presence in Macedonia throughout the Balkan Peninsula. This had been brought home to GE on his fleeting visit to Serbia, which had achieved independence from the Turks in 1877. Here, on what was once the periphery of the failing empire, he described his first impressions of the notorious fortress of Belgrade. The fortress, a stark reminder of Ottoman domination, was reconstructed by the Turks in the late nineteenth century. Built on a hill at the apex of the town, in a triangle formed by the Sava and the Danube, it is steeped in history:

> It is quite mediaeval and Turkish but is still garrisoned – chiefly it seems by convicts with leg chains who work in gangs. The Danube here breaks up among hills and islands and the view is superb. In a strange way the

> fortress seems to mark the end of things and to still be the outpost of the Ottoman world defying all Europe proper.

Breaking down the barriers between England and the Ottoman Empire was, of course, the original *raison d'être* of the Levant Consular Service. The service was inherited from the Levant Company, an English chartered company formed in 1581 to regulate English trade with Turkey and the Levant. The company had, in many respects, provided the model for the better known East India Company, although it never sought territorial control. In 1592 it was granted a monopoly over the lucrative trade in the eastern Mediterranean.

Membership was open to any merchant over the age of 26 wishing to trade with the Ottoman territories on payment of a fee of £25. The company would in return represent the interests of those merchants throughout the empire, where special privileges conferred by the sultan smoothed the way to easy profits. When the Levant Company's charter was revoked in 1825 its consular service was taken over by the Foreign Office.

By the early twentieth century, European consuls in the Levant no longer had the level of influence enjoyed by their predecessors. However, they still had a significant role to play. Levant consular posts were, according to GE, 'political observation posts for seeing and reporting remote parts of the Near East, which up to the time of the First World War was the "powder magazine" and chief danger spot of Europe'. It was, he recalls, an interesting and responsible job, though often a lonely and somewhat dangerous existence which resulted in 'a rather alarmingly high level of suicides, candidates for lunatic asylums and victims of murder, riots and brigands'.

The consuls also had a more mundane role. Under what were known as the Capitulations, British nationals in the Ottoman Empire were granted a number of privileges including exemption from local taxes and local prosecution. They could instead bring their disputes before the local consul under the Common Law of England and Wales. Consuls were, in effect, 'judges, marriage officers, registrars, postmasters, coroners, Commissioners for Oaths and in fact every kind of official rolled into one'.

At the time GE arrived in Sofia, however, a volatile political situation and general British distrust of the Bulgarians meant there were few British nationals needing his services. The consulate was a diplomatic mission in all but name. Despite being a monarchy the country was still predominantly the same peasant culture that had existed under the Ottomans more than fifty years earlier. GE recalled that the king, 'imported from Coburg', gave his ministers 'who were all sons and grandsons of serfs from the days of Turkish rule' a terrible time. If the king wasn't out for your blood your political enemies were.

The internal settling of scores was secondary, however, to the wider regional issue of Macedonia that occupied most of GE's working day. Telegrams arrived on a daily basis 'with accounts of fresh outrages on the

Bulgarian peasants and reprisals on the Turks'. GE's prediction that war was 'quite on the cards' was realised the following year when the united forces of the Balkan League drove Turkey out of Europe for good.

At a social level GE found that consular life in Sofia was too exclusively diplomatic for his taste. The society of unadulterated diplomats was 'rather trying, especially as I had to dine with them nightly at the club'. More pleasant by far was to walk out to the villages at the foot of the Witosche – the big mountain that dominates Sofia – to go on an (unproductive) bear hunt or to discuss politics in Turkish with older Bulgarians who had learned to speak the language under Ottoman rule.

GE's next port of call was Canea in Crete, where he was to stand in for the Vice-Consul who was on leave. This, too, was a mainly political posting but the situation was less clear cut than in Bulgaria. Crete was essentially a 'no man's island' as regards possession. Having wrested control from the last Turkish sultan after a series of massacres, the Great Powers had no desire to hand it back to the Greek 'mother country'. As a result it was in a state of political limbo and acute unrest.

GE hadn't been in post for long before he began to worry about the lack of opportunities in the consular service. By 1911, now a few months into his post in Crete, he was already considering reading for the bar. He recalled that Albert Wratislaw, who was then Consul-General in Canea, told him that

> it would be an excellent thing to get 'called'. Most of the really successful men in our service are out of it, so to speak, i.e. they have left it for much better jobs later on. For instance Sir Adam Block is now English Representative on the Ottoman Debt and draws a salary of over £3000.

A salary of £3,000 in 1911 was more than three times the amount paid at the top of the Levant Consular Service. GE was aiming high, even for a young man with a 'more than average amount' of ambition. However, his remarks shouldn't obscure the fact that he had genuine respect for his superiors in the service. He had nothing but admiration for Wratislaw himself and he fiercely defended Arthur Alban, his next chief in Cairo, against members of the diplomatic service who considered themselves to be superior in ability and social status. Alban, who he described as 'a man of integrity, common-sense, careful and fair judgement and human kindness' was 'treated like mud by the diplomatic snobs at the Residency'.

There seems to have been well-founded concern, shared by other young recruits into the service, that their long and demanding training could indeed come to nothing if jobs dried up as the Ottoman Empire shrank. In his biography of the service D. C. M. Platt noted that by 1914 the experience on average was that it took about 16 years before a man was promoted to consul on a salary of £600 per annum, leaving nothing to spare for a married man and a family.

GE was still taking exams when he arrived in Cairo, two years after his first assignment and four years after the open competition. All student interpreters were expected to take further examinations in the civil, criminal and commercial law of Turkey as well as the history, language and administration of the country in which they served: a seemingly demanding requirement for a job which, on the lower rung of the promotional ladder, involved a daily routine that made relatively few demands on the intellect.

In contrast to his previous postings Egypt was all 'glare, heat and jostling crowds'. There weren't enough hours in the day to deal with the stream of visitors demanding his services. Some were more deserving than others. One day a man had walked into the consulate dressed like a boy scout. He claimed he had just walked from Cape Town and needed his fare home as he'd been robbed in the Congo of everything except his shirt. Informed that this was a bit too much to ask, he was given a bed for the night and then sent on his way. Other visitors included:

> an ex-Sultan of the Maldives (we had ex-ed him some years ago for ill-conduct generally) and a troup of girl artistes from a Cairo music-hall. I gave an Englishman a passport to Mecca some two or three months ago (he said he was a Moslem going on a pilgrimage); he was captured, mortally-wounded and wearing a Turkish uniform, by the Italians in Tripoli a fortnight ago. There seem to be a fair number of Englishmen in the Turkish army.

Two weeks later GE outlined a further day's work which,

> if somewhat unintelligent, has at least the merit of variety.
>
> 0930 Issuing passports to a frowsy Maltese family and (admonishing) the materfamilias in my best Italian for various shortcomings
> 1000 Criminal case in court for assaulting a 'chawish' [police officer] on duty. After the usual perjury [. . .] the defendant is sent to spend a summer *villégiature* [vacation] of 2 months in gaol
> 1100 Marrying Mr Luigi Briffa and Miss Cesira Cumbo [. . .] with some difficulty as neither of them knows a word of English
> 1200 I drive with Alban up to the Citadel to 'view the corpse' of a poor gunner who died yesterday after drinking lemonade – possibly poisoned. We sign his death certificate and arrange for an inquest

Then to the Club for lunch. In some ways GE was well suited to this aspect of the job. Somewhat reserved by nature, but with an unrelenting interest in the world about him and an assiduous attention to detail, he was perhaps the ideal civil servant. But he had set his sights higher than this.

In Cairo he began to make enquiries about switching to the Egyptian civil service, which was run by the British under a de facto protectorate. In the short term, however, there were other, more personal, pressures to contend with.

* * *

Visits home were few, and frequently organised to suit the demands of the service rather than the wishes of the employees. Short postings, particularly in the early years, made it difficult to establish close relationships. At times, GE confessed, a feeling of loneliness and alienation lay heavily on him: 'it seems such a hard thing that one should spend one's life right away from the people one loves best and the country which is one's own'. But that was soon to change in one respect at least. Some ten years earlier, when living in Gotha, GE had met a young girl who was staying with her family in a *pension* for English visitors. A few years later, when undergraduates at different universities, their paths crossed again: this time in St Jean de Luz in France, where his mother now ran a school. The girl's name was Dorothy Johnson.

From that second meeting onwards, GE recalls, he was 'tangled in her net' and 'we started a postal exchange which never flagged during the long intervals which were to occur between our later meetings'. The relationship survived the long intervals and in October 1912, on one of GE's visits home, they finally announced their engagement. GE was overjoyed. Now, at last, he would have someone to share with him the joys and frustrations of colonial life: frustrations that had a tendency to 'develop into real melancholy' when on his own.

The period of waiting was not over yet. It would be more than a year before they would see each other again and maybe more before they could be married. Risking the disapproval of the extended family, they decided to limit the endless round of social visits that normally accompanied an engagement. As GE explained: 'Our time together is so short and the gap ahead so big' that this entrenched middle-class convention was politely by-passed in favour of a few precious weeks together.

The decision to get married at all had been a difficult one. Employment in the service in the early twentieth century was considered secure, and was therefore much sought after, but early marriage was discouraged. A letter in the Foreign Office archives dated 1906 states that 'as junior officers in the Levant Service may be required to proceed at short notice to any part of the Ottoman dominions, Egypt, Persia or Morocco it is most undesirable that they be married.'[1]

In addition it was pointed out that a Vice-Consul's salary of £300 a year was only sufficient to maintain a single man in the required manner. Following the Ottoman defeat in the First Balkan War of 1912, the outlook for young men like GE seemed grim. Would the service contract along with its original *raison d'être* and what chance did he have of progressing through the ranks? Now Acting Vice-Consul in Cairo, he doubted that he and Dorothy could survive on the money he would receive even when he got a permanent post.

He was not alone. 'The pessimism that pervades the service as a whole' was a matter of widespread concern according to the second dragoman (interpreter) at the embassy in Constantinople, who told a 1912 Royal Commission on the civil service that 'on pay that was no better than that in the home civil service, they had to endure "a lifelong expatriation" in posts that were for the most part remote, primitive and unhealthy.'[2]

Back in Cairo after announcing his engagement, the question of finance weighed heavily on GE. Dorothy's father, a Brigadier-General in the Royal Artillery, wanted to know what family money might be forthcoming if his income was insufficient to support a wife – an intrusion into the personal sphere that would be unthinkable today. But it wasn't just GE's finances that were to come under scrutiny. The Brigadier-General also suggested that he make some sort of settlement on Dorothy in the event that there was some 'change of affection' on his part – a suggestion dismissed by GE as 'rot' and 'a contingency we need not, I think, contemplate!'

In September 1913 all these concerns were brought into sharp focus for GE. The Foreign Office announced his appointment as Secretary to the Boundary Commission and also stated that, on its conclusion, he was to take up the post of Vice-Consul in Mosul. His delight at landing the Commission job was unqualified and he was relieved that the years of uncertainty were behind him. But he had already decided that Mosul was probably not the ideal place for Dorothy to get her first taste of consular life. 'It is conceivable', he had written in an earlier letter home,

> that I might be sent to Mosul – on the Tigris, a most barbarous and out of the way place, but interesting in some ways. You get there by floating down the Tigris on skins, but as you can't float up again I don't know how you return.

It was, as he now wrote in a rare letter to his father 'an interesting place but not the place I should choose for starting married life at!'

Choice, however, was a luxury that he didn't have. On 16 October 1913, less than five weeks after receiving the telegram confirming his appointment, he married Dorothy at Hayes Parish Church in Middlesex. On 21 November, after a honeymoon spent travelling through France, he left his new wife on the quayside at Marseilles as he boarded the P & O ferry to Port Said on the first leg of his journey to the Persian Gulf. It was a moment of mixed emotions for GE. I have no record of how Dorothy was feeling. Sadly, no letters to Ernest have survived. I can only conclude that he felt no need to keep them once the period of separation had ended. When he left to take up his new job

2. Wedding photograph

just one month after his wedding, however, that moment seemed a long way away.

* * *

GE was no stranger to painful farewells. They had been a regular feature of his childhood, and members of the extended Hubbard family had been living and working abroad for generations. In 1769 his great great grandfather, William Hubbard, emigrated to St Petersburg and set up an import/export business between England and Russia. The business did well, despite having to close during the Napoleonic wars, and was re-launched by his two sons when peace was restored.

In 1842 the next generation turned to the manufacture of cotton. Still in St Petersburg, they built a spinning mill and a weaving mill in quick succession.

A derelict dyeing and print works was bought and renovated in 1866. The import/export business was gradually dropped as the firm opened agencies for the sale of its cotton products throughout the Russian Empire. At the same time it diversified further, buying timber concessions and a saw mill on the Onega River. When these concessions ran out, further investments were made into forests in Finland.

Throughout the mid- to late nineteenth century the family prospered. GE's grandfather, William Egerton, ran the mills and print works in Russia. His great uncle, John Gellibrand, later the first Lord Addington, ran the partner company in London. He was also a Tory MP and subsequently became Governor of the Bank of England. In the words of Martin Daunton, the family

> seemed by the mid-nineteenth century to have emerged as a City dynasty [. . .] firmly entrenched in the Bank of England, leading supporters of the Conservative party, significant landowners and recruits to the aristocracy by marriage and ennoblement.[3]

For many of the young men in the family a spell in Russia after finishing their formal education was a rite of passage. GE's older brother, Leonard, went to work in the firm's Moscow office after leaving Cambridge in 1903. But by then, however, the bubble had burst and the Russian businesses, now run by GE's uncle and cousin, were in serious difficulties.

A loan of some £90,000 was secured from City and family contacts and immediate catastrophe was averted. The businesses struggled on with the help of more loans but, due to poor leadership and a failure to manage issues of inheritance, they never really recovered. Family members who had been dissuaded from withdrawing their capital lost large amounts of money. The losses were compounded, says Martin Daunton, because 'the firm continued to have access to capital, but its poor performance meant that both firm and family failed'.[4]

Leonardslee, the country estate that GE's grandfather bought in 1853, had been a proud symbol of the achievements of the Hubbard enterprise abroad. Set in a thousand acres of beautiful Sussex countryside, it retained the promise of future success throughout the period of financial collapse. In 1889, when Cyril's older brother could no longer afford to maintain it, he sold it to his brother-in-law, who kept up all the family traditions. A young child visiting Leonardslee, as GE did on numerous occasions, would have been unaware of the drama playing out in the background. It would have been an environment to encourage confidence and self-belief in all but the faint-hearted.

GE later reflected in his unpublished memoir, written in the 1940s, that 'the feeling one got in a big house or establishment like Leonardslee, with its 20 gardeners, gamekeepers, home-farm, hunters, hacks, carriage-horses and ritualistic daily life was of something that would go on for ever'. Although now convinced that it was 'all wrong' that so much of the country's wealth should have been used up for pleasure by a small class of people, he was equally insistent that 'it had its dignity, a certain beauty and some goodness in the friendly and solid relationships between "master" and "servant"'.

At one level GE's immediate family was only on the periphery of the wider family crisis. But it, too, was dependent on the family fortunes. Cyril, as a younger son, had never been involved in running the business but his inherited capital had brought him a life of some comfort. The census of 1901 showed the family living in Hartfield with a cook, a nanny and four housemaids. The sudden loss of the income that had allowed him to live above the modest level of a vicar's stipend had been the driving force behind the move to Gotha later that year. This was the point at which, with Leonard at Cambridge and five younger children to support, GE's mother Agnes 'had to take the job of earning for our education'.

It was the Hubbards who provided the worldliness and sometime wealth in GE's life but it was his mother's side – the Scotts – who provided the stability and 'simple but solid comfort' that he held so dear. Agnes was herself the daughter of a vicar, Thomas Scott, who had worked for many years in London's East End. She was GE's lifeline to home throughout his travels in Europe and it was from her that he inherited his passion for knowledge and his unremitting work ethic. Indeed it is only from his letters to her, carefully stored and cherished over the years, that we have so full a picture of his early life.

GE was 16 when the Hubbard income dried up and he had seven more years of study before taking the consular service entry exam. His mother's contribution to the family finances was crucial and, when Cyril took a job in St Jean de Luz in 1903, Agnes moved her girls' school into their new home. Four years later his father, aged just 53, suffered a mild heart attack whilst working as a locum in St Moritz to pay for the family summer holiday.

As Cyril's health became progressively worse and it became clear that the capital from Russia would never materialise, Agnes decided that her school would need to be expanded. Paris, it was decided, would be the best place to do this. So in his last summer in England GE joined her in Paris to hunt for 'an old chateau or some similar suitable site for a school in the environs of the city'. Their mission accomplished, he returned to Cambridge for the last time before leaving for Constantinople. Agnes successfully ran the Paris school for 25–30 girls until forced to leave the country at the outbreak of World War I.

The family's struggle to give GE and his siblings the best possible education despite their reduced circumstances throws some light on GE's seeming obsession with money in his early years in the consular service. Even then, still owing the £500 bond demanded by the service, he had only partially thrown off his dependence. The appointment as Secretary to the Boundary Commission was not only a job that exceeded all his expectations: it also marked the moment – at the age of 28 – that he became, at last, financially independent.

* * *

The Boundary Commission appointment had come out of the blue. GE's 'old chief' in Crete, Albert Wratislaw, had been chosen to lead the British delegation. He had put GE's name forward and wrote to ask him if he would accept. A Foreign Office telegram followed. He would be formally appointed Vice-Consul to Mosul before leaving for the Gulf. He would then take up his duties in Mosul when the Boundary Commission had finished its work. There was no indication of when, if at all, he would get any leave. It threw his marriage plans into confusion but it was, of course, irresistible. He accepted.

When he left Marseilles after his honeymoon on 21 November 1913 GE had little idea of what lay ahead. That would not become clear until he reached the expedition's 'base camp' in Mohammerah (now Khorramshah) on the Persian Gulf. In the meantime, however, the first leg of the three week journey via the Suez Canal, Aden and Bombay was a sociable affair. Travelling in a group with other delegation members they soon got to know their fellow passengers and spent their days playing bridge, dancing and devising ingenious games to pass the time. As they drew nearer to the Gulf, however, discussion turned to the task ahead: how to avoid the unbearable summer heat, whether the survey work could be done in the time allowed, and the problems that could arise where the border was still in dispute.

More immediately, however, the challenges facing them on arrival in Mohammerah were practical ones. It was fortunate, therefore, that the delay caused by Turks and the Persians – in no hurry to arrive or to move on when they did – gave them two whole months of preparation before the expedition finally set out.

If there were briefings about the nature of his job in the few weeks before his departure from Marseilles then I have no record of them. There certainly seems to have been little or nothing provided in the way of physical preparation for this journey of '1200 miles on horseback or foot across difficult and possibly dangerous country.' Though clearly aware of the hazardous nature of the task ahead, GE had total confidence in the delegation leaders. In addition to Wratislaw, of whom he said 'I ask for no-one better to work for', there was also the invaluable experience of Deputy Commissioner Captain Arnold Wilson to draw on.

Wilson, later to become the Acting Civil Commissioner in Baghdad, was an officer in the Indian Political Service and a former Consul in Mohammerah. In a later, posthumously published, record of his years as a political officer he recalls how he made the bodily comforts of all the Commission delegates his first priority:

> I spent a week pitching the tents in the desert on carefully spitlocked lines, based on my recollection of the Royal Camp at Rawalpindi upon which I had worked with my regiment when King George V as Prince of Wales visited India in 1905. It was lit by electric light furnished from

> the Oil Company's plant; patrolled by guards furnished by the Shaikh of Mohammerah.[5]

The Turks, apparently, said the tents were too large, the Persians they were too small, the Russians that they were too cold. To GE they were luxurious beyond his wildest dreams:

> People say we are more luxuriously housed than the high officials at the 'court of Delhi' [. . .] they are about the size of a cathedral and each takes five mules to carry it! Camp will be about as uncomfortable as the Ritz I should say.

GE's own tent had two doors in the front, with venetian blinds to use when the flaps were up. These opened onto a porch that bore his name plate. The living room had a floor space of about 13 square metres and was 3 metres high. It was furnished with armchairs, a camp table, two yakdans (trunks), and more on offer if required from the lavish camp stores. There was also a separate bathroom and bedroom.

GE was to appreciate the seemingly extravagant living accommodation as he struggled with the elements and all the other travails that beset the expedition in the following months. At the moment, though, there was another equally important task ahead – that of choosing the horse that would be his daily companion from the Gulf to Ararat. The challenge was to find the animal best suited to carry him across the scorched deserts of Mesopotamia and the often treacherous mountains of Kurdistan.

Thanks in part to his time spent in Rennes, where he had 'managed to get riding lessons in a military "manège" [riding academy] where stirrups were taboo and most of the riding was bareback' he was already a skilled horseman. But it's unlikely that he had ever before had to 'ride girth deep through water and mud' or 'come down a breakneck track less than a foot broad in the dark'. He needed to find a horse of exceptional quality to take him through these uncharted waters.

On arrival in camp GE had discovered that a horse he didn't much like had already been bought for him. This had to be remedied and it wasn't long before he found the opportunity to do so. The delegation members had been invited to spend a day with employees of the Anglo-Persian Oil Company, who formed the bulk of the British colony in the area. During the visit GE rode a horse belonging to one of the men whom, he discovered, was looking to sell it before departing for a spell of home leave. GE knew immediately that this was the horse he wanted. He persuaded Wilson to buy it for the Commission so that he could then 'bag him instead of the "pony" already bought'.

GE's plan appears to have worked, for this horse was to be his trusty companion throughout the expedition. Christened Archibald by its new owner, it had quite possibly been a 'racer'. It was certainly hot-headed and inclined to the chase. GE had work on his hands as some training was clearly required. By

3. G.E. Hubbard on Archibald

19 January he cautiously observed that he 'is improving somewhat but [. . .] our opinions frequently differ as to the pace and direction we shall go in'. By the middle of February, when they finally set off on the first leg across the desert, he was able to report that: 'Archibald was perfect – a trifle over-keen but that will wear off, and he has the most comfortable walk of any horse I have ridden.'

Having found Archibald, GE had soon discovered that all other practical considerations could be left in the able hands of the Deputy Commissioner. Captain Wilson was clearly a master organiser and the other delegations, despite their express dissatisfaction with the size and temperature of the tents, were not slow in availing themselves of his services. The captain recalls that

> within a month the three delegations asked me to supply them with all the fodder and grain needed for their escorts and [. . .] not long after I was asked to provide them with all their requirements of food so far as it could be locally purchased.[6]

There's little question that Wilson's skills in this department were invaluable in unifying the delegations; as, too, was his chosen method of oiling the wheels of diplomacy when matters threatened to get out of hand. As a

4. The Persian Commissioner and the Turkish Commissioner (fourth and fifth from left)

result 'our proceedings were remarkably harmonious. Alcohol is a valuable industrial solvent; it was not less useful in our particular field'.[7]

With Wratislaw and Wilson taking care of the wider needs of the Commission, GE was free to concentrate on his own job. There was indeed a consensus that relations between the delegations were surprisingly harmonious although, as GE was to discover, the going was not always easy. There were to have been four Secretaries on the Commission – one for each delegation. In the event there were only three, the Russians having arrived without one. At their first meeting in a Persian house in Mohammerah on 19 January 1914, GE sat between his Turkish and Persian colleagues.

> The Turk on my left was fearfully business-like and took down every word verbatim – utterly unnecessary – while the Persian on my right sat stolidly doing nothing until half way through he turned to me and said 'What am I supposed to be doing, do you know?'

GE doesn't record his reply but it's clear that he, like other members of the British delegation, decided to assume a leadership role from the start. 'I'm going to be rather a buffer state', he wrote home after that first meeting, 'for the Turk, though very pleasant to me, simply ignores our Persian colleague or else is rude to him.' Hardly surprising, when one remembers that this Commission had been set up to settle disputes that the Turks and the Persians had been unable to resolve for more than three-quarters of a century. Nevertheless, equilibrium was maintained and, as far as we know, the other two members of the Secretariat had no objections to GE taking the lead in what was to be a painstaking job.

As GE explained, 'The result of each meeting and act of the Commission have to be recorded accurately in the *procès-verbal* and each list of the frontier as it is settled must be exactly described besides being marked on the big maps.' The records of the meeting were all typed up in English and French. French had historically been the language of diplomacy in Europe from the eighteenth century onwards and was subsequently exported further afield by Napoleon. It was still the most spoken international language in the Levant in the early twentieth century. It was also, given its ability to convey subtle shades of meaning, particularly suited to the task ahead. It was a task that demanded total fluency, as the final report, or *procès-verbal*, was to be incorporated into the November 1913 Protocol as the only official record of the Commission's decisions.

It was, as GE was to discover, a mammoth task and one that occasionally threatened to overwhelm him. By the middle of March he was finding that he 'had to do the work of the whole of the Secretariat'. His Turkish and Persian colleagues, he complained, hadn't even brought typewriters, so he had to supply all the Commission with documents as well as write most of them. This was confirmed by Captain Wilson, who noted that: 'Wratislaw and Hubbard do all the diplomatic work.'[8]

Despite his complaints, however, GE was not altogether unhappy that the work of the Secretariat was 'more or less concentrated in my own hands'. The single-mindedness that had taken him through his unconventional teenage years was unquestionably his strongest characteristic. A complex man, not physically strong, he was by his own admission a bit of a loner. This, rather than hold him back, only reinforced his determination to make a success of everything he embarked upon.

There was little doubt amongst his colleagues that GE was an indispensable member of the British delegation. Without his records the Boundary Commission's deliberations would never have seen the light of day. However, this book is not primarily about the Commission. It is about people not politics. It is a snapshot of the fascinating and diverse mix of cultures that have inhabited this compelling but unforgiving terrain over the centuries. It is a story of peoples driven more by the forces of nature than the will of man; nomads ebbing and flowing across lands that defy demarcation by lines on a map – lines that then, as now, seemed as though they were made to be breached.

The international politics of another era hovers inescapably in the background, only forcing its way into the narrative in the last few weeks of the expedition. But it was, it appears, local tribal politics that were responsible for the misfortune that befell GE shortly before his journey's end. That, however, is another story – one I will return to in the *Afterword* when GE's tale ends.First I leave you to join him on his journey: a journey that I hope will give a feeling for a landscape and a way of life that is too often lost in the telling of the history of this troubled land.

Notes

1 National Archives, FO 369//52/28 File 14975. Asks whether Student Interpeters in Levant should be unmarried and whether the salary is sufficient.

2 G. R. Berridge (2009), *British Diplomacy in Turkey, 1583 to the Present: A Study in the Evolution of the Resident Embassy*, Martinus Nijhoff Leiden, p. 94.

3 Martin J. Daunton (2008), *State and Market in Victorian Britain*, Boydell Press, Woodbridge, pp. 182–3.

4 Ibid., p. 168.

5 Arnold Talbot Wilson (1942), *S. W. Persia: Letters and Diary of a Young Political Officer 1907–1914*, Readers Union, London, pp. 272–3.

6 Ibid., p. 273.

7 Ibid., p. 275.

8 Ibid., p. 280.

Note

All direct quotes, unless otherwise indicated, are taken from G.E. Hubbard's personal letters and memoirs. I can take no responsibility for all the statements made in these papers. However they have, wherever possible, been underpinned by other records in the family archive or by independent outside sources. The Hubbard family archive is in private ownership.

Bibliography

Berridge, G. R., *British Diplomacy in Turkey, 1583 to the Present: A Study in the Evolution of the Resident Embassy* (Leiden, 2009).

Crampton, Richard, *Bulgaria 1878–1918: A History* (New York, 1983).

Daunton, Martin J., *State and Market in Victorian Britain* (Woodbridge, 2008).

Epstein, M., *The Early History of the Levant Company* (London, 1908).

Mansel, Philip, *Levant: Splendour and Catastrophe on the Mediterranean* (London, 2010).

Platt, D. C. M., *The Cinderella Service: British Consuls since 1825* (London, 1971).

Wilson, Arnold Talbot, *S. W. Persia: Letters and Diary of a Young Political Officer 1907–1914* (London, 1942).

PREFACE.

I HAVE to preface this book with an apology—and, worse still, an excuse. The apology is for venturing to publish at the present time the record of a journey which (except for quite the last stage) took place in that almost prehistoric epoch before the war. My excuse is twofold: firstly, that fifteen months of enforced idleness drove me into writing it; and, secondly, that subsequent events have contrived to add a special interest which it could not otherwise have claimed. Although in no sense a "war-book," it deals with countries which have been the scene of two, if not three, campaigns in the present war, and on this fact I rely to justify my temerity.

The first of these campaigns—taking them in the order of our journey from South to North—is the British Expedition to Mesopotamia. Its main features are so well known as barely to need repetition: the landing of our force at the mouth of the Shatt-el-Arab in November 1914; the battle of Fao and the fall of Basra; the expedition under General Gorringe up the Karûn to

Ahwaz to safeguard the oil-fields and the pipe-line, ending in the successful "rounding-up" of the Turks at Amara; the storming of Kurna, the advance up the Tigris, the pitched battles at Kut-el-Amara and Ctesiphon, and, last of all, the gallant stand and ultimate surrender of General Townshend's force. But unless one has been to the country, seen the desert and the marsh and the date-groves lining the Tigris, and known—even in its mildest form—the heat of those limitless plains, it is impossible to conjure up any true picture of what our British and Indian troops went through, or fully to realise their extraordinary fortitude. I hope, therefore, to supply in the earlier chapters of my book a slight background for this campaign.

Apart from the actual fighting, moreover, a certain degree of interest must centre around the town of Basra, the port of Mesopotamia and the one important place which we occupy and administer. It is an interest, too, which is hardly likely to vanish with the end of the war. Without hazarding any rash guesses into the future, one may well recall in this respect the words of the Viceroy of India to the people of Basra, spoken during his visit to that town in January 1915: "The British occupation has raised problems which require prompt consideration and settlement. I have come here to see local conditions for myself in order the better to judge what measures are necessary. You are aware that we are not engaged single-handed in this great struggle, and we cannot lay down plans for the future without a full exchange of views with the other great Powers, but I can hold out the assurance that the future will bring you a more benign rule."

Coming next to Bagdad, Kasr-i-Sherin, and the Kermanshah road, which form the subject of Chapters VII. and IX., we enter the sphere of quite another series of operations. This was the route by which the Turkish troops entered Persia last year, hoping by joining hands with the German-led Persian rebels to wreck the influence of England and Russia in that country; only to retreat again by the same road after successive defeats by the Russians at Hamadan, Kangaver, and Kermanshah. At the time of writing, General Baratoff's troops have reached and taken Kasr-i-Sherin itself.

Finally, the Northern part—that is, Azerbaijan. The details of the fighting there are probably less familiar to people in England, but it will be remembered how in the autumn of 1914 the Turks, who were just then making a bold attempt to reach Tiflis, violated the Persian frontier near Urmia. The story of the wholesale massacre of the Christian (Nestorian) population by the Kurdish irregulars—who ravaged as far as Tabriz, burning hundreds of villages and driving the weak Russian garrison out of the province for the time being—reached us only in scanty paragraphs and an occasional letter or two in the Press. Of the events in the immediate neighbourhood of Urmia, I give a brief account in the last chapter of the book. As the tide slowly turned against Turkey in the Caucasus, and the Russians advanced towards Trebizond and Van, the invaders of Azerbaijan found their position untenable, and by the spring they were back again on their own side of the frontier. Thus have the tentacles of the great world-war reached out to and embraced practically the whole of that remote region of Western Asia which was the scene of the journey described in the following pages.

The substance of this book consists of little more than a record of personal experiences and impressions of the tribes and countries through which we passed. Politics lie outside its scope, and I have condensed within the limits of a single chapter what little I have to say on such general topics as the connection between our own country and Mesopotamia. Even the thrilling adventures which season so many travellers' tales play, I fear, but a small part in my narrative, and the utmost that I can confidently promise the reader is to conduct him (should his patience permit) by little-trodden paths "from the Gulf to Ararat."

MAY 1916.

NOTE.

THE majority of the photographs which illustrate this book were taken by Captain Brooke, who has kindly allowed me to use them for this purpose; for a great number of the remainder I am indebted to Mr Wratislaw.

Captain Wilson has been of the greatest help to me in many ways, in particular by lending me his notes on various districts.

Finally, I have to thank Mr W. Foster, C.I.E., of the India Office, for enabling me to consult extracts from the early records of the Honourable East India Company, and the Editor of 'The Near East' for the loan of old files dealing with irrigation in Mesopotamia.

I have acknowledged assistance from other writers in the body of the book.

G. E. H.

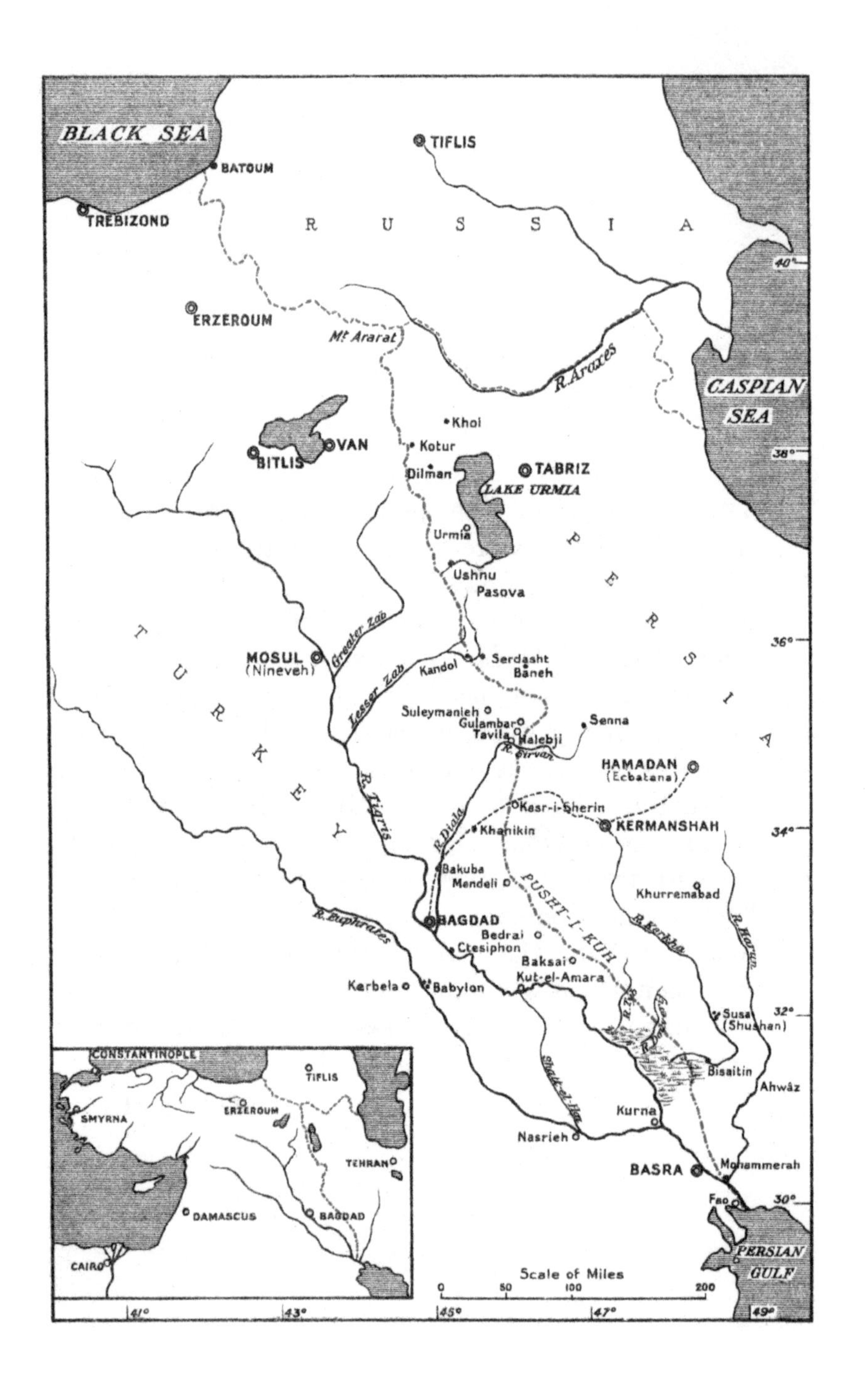

BLACK SEA
BATOUM
TIFLIS
TREBIZOND
R U S S I A
ERZEROUM
Mt Ararat
R. Araxes
CASPIAN SEA
Khoi
VAN
Kotur
BITLIS
Dilman
TABRIZ
LAKE URMIA
Urmia
Ushnu
Pasova
P E R S I A
T U R K E Y
Greater Zab
MOSUL
(Nineveh)
Lesser Zab
Kandol
Serdasht
Baneh
Suleymanieh
Gulambar
Tavila
Halebji
Senna
R. Sirvan
HAMADAN
(Ecbatana)
Kasr-i-Sherin
KERMANSHAH
Khanikin
R. Diala
R. Tigris
Bakuba
Mendeli
PUSHT-I-KUH
Khurremabad
BAGDAD
R. Euphrates
Bedrai
Ctesiphon
Baksai
R. Kerkha
R. Karun
Kut-el-Amara
Kerbela
Babylon
Susa
(Shushan)
Bisaitin
Shatt-el-Hai
Ahwâz
Kurna
Nasrieh
BASRA
Mohammerah
Fao
PERSIAN GULF
CONSTANTINOPLE
SMYRNA
ERZEROUM
TIFLIS
TEHRAN
DAMASCUS
BAGDAD
CAIRO
Scale of Miles
0
50
100
200
40°
38°
36°
34°
32°
30°
41°
43°
45°
47°
49°

FROM THE GULF TO ARARAT.

CHAPTER I.

THE FRONTIER.

In the autumn of 1914, a few weeks after the outbreak of the War, a small party consisting of Englishmen, Russians, Turks, and Persians arrived on a snow-clad spur of Mount Ararat, constructed an unobtrusive stone pillar beneath the shadow of the great 17,000-foot peak, and swiftly dispersed, each to his own country. Such (thanks to the Hun) were the inglorious circumstances which attended the culminating act of seventy odd years of diplomatic *pourparlers*, special commissions, and international conferences between the four Powers concerned.

To give more than the barest outline of the history of the Turco-Persian frontier question during those seventy years is manifestly out of the question; the dossier containing its record being, I verily believe, of a length to stretch from end to end of the entire 1180 miles which constitute the actual frontier. At the same time, a short summary is, I think, excusable, firstly, in order to explain the origin of our journey; secondly, as an example (though happily not a very typical one) of the way a frontier comes into being; and thirdly, as a phenomenon of procrastination unparalleled even in the chronicles of Oriental diplomacy. Fate has contrived, moreover, to stage this particular scrap of history in her most dramatic style by arranging a "curtain" for each one of the three principal acts in the shape of a European war—the Crimean, the Russo-Turkish, and lastly the Armageddon of to-day.

You may trace on a map of Western Asia a dark line of mountains which, after disentangling itself from the confused mountain-mass of Armenia around Ararat, stretches S.S.E. to near the top right-hand corner of the Persian Gulf. On the map it looks, perhaps, rather a smudge than a line, for it spreads out to a great breadth in a series of parallel ranges. But the backbone or "chaine magistrale"—to borrow an expressive phrase—is in reality fairly clearly defined, and constitutes a huge barrier, difficult to cross in summer and impassable in winter, separating the great Persian plateau from the plains of Mesopotamia. It is, in fact—like the Alps or the Pyrenees, or, on a far larger scale, the Himalayas—one of Nature's frontiers, and nations have conformed to it as such from the earliest historical times. It served as a boundary on the East for Assyria, and on the West for Media. Often, of course, the countries on either side have been temporarily united within a single empire, as, for instance, when Alexander from the one side thrust through to India, or Khosroe from the other extended his dominions to the shores of the Mediterranean, or again when a human avalanche under Chengiz Khan or Tamerlane engulfed everything, and swept all frontiers clean out of existence. But sooner or later this great mountain range has invariably resumed its original *rôle* as a divider of nations, and in this capacity it has always formed the basis of the frontier negotiations between the Governments of Turkey and Persia.

These two Empires first came into contact early in the sixteenth century. At that time Persia was rapidly recovering from the devastation of the Mongol invasions, and finding again her entity as a nation. Her *risorgimento* happened to correspond with the amazing wave of success which carried the Ottoman armies as far afield as Egypt and Tunis, and, eventually, up to the very gates of Vienna. Suleyman the Magnificent, turning his attention momentarily towards the East, conquered Northern Kurdistan and Azerbaijan, and marched his armies down to Bagdad, which he captured in the year 1534. Then the tide turned in favour of the Persians. Shah Abbas—their Charlemagne—retook most of the Northern provinces, expelled the Turks from Bagdad, and established a frontier along the Tigris. The pendulum having thus swung both ways, finally came to rest in 1639, when Murad IV. definitely won Bagdad for Turkey, and settled the frontier by treaty along very much the same line as it follows to-day.

It was almost exactly two hundred years later that England and Russia first became involved in the Turco-Persian frontier dispute. Murad's treaty had by then disappeared, the Turkish original having been destroyed in one of the big fires which are such frequent events at Stamboul, the Persian in one of the revolutionary upheavals of hardly rarer occurrence at the Persian capital. In 1842 frontier frays on a large scale brought the two countries to the brink of war. The Persians complained of an unprovoked attack by the Turkish ruler of Suleymania on their subject, the Vali of Ardelân, while the Turks retorted that the fertile district of Zohab, assured to them by the treaty of 1639, was still occupied and held by Persia. Another incident which had helped to bring matters to a crisis was an attack on Mohammerah by the Pasha of Bagdad, an account of which last exploit from the diary of General Tchirikoff is perhaps worth quoting as a sample of Arab warfare:—

"Ali Riza Pasha led his force from Bagdad down the right bank of the Tigris, conveying his provisions and supplies in boats.

"For the defence of Mohammerah the whole Chaab tribe assembled to the number of about 25,000, as is asserted with evident exaggeration. The town was filled with them, those who could not find room in it occupying a piece of ground on the left between the palms and the desert.

"Ali Riza Pasha opened fire on the town with his artillery without delay. The Arabs replied from the wall with their gun, and with fire from small arms. But the fire of neither side carried far enough. Ali Riza Pasha went to his artillerymen, reproved them for firing so badly and for not putting enough powder in the guns, and promised 5000 piastres to any one who knocked over the gun in the town. Soon afterwards the Mohammerah gun burst, probably owing to the inexperience of the Arabs in regard to artillery, and to overcharging. Several men near the gun were killed. The Turks thought their fire had destroyed the gun, and they even name a young artilleryman, who, they say, fired a ball into a ball which had not yet left the Arab gun, and thus caused the latter to burst. This event produced a panic among the Arabs. The Turks then made an attack in three divisions, the artillery in the centre, one regiment of infantry on the right, the other, with the Muntefik, on the left.

"The defenders of the town all ran away before the Turks reached the walls. Thus the town was taken without fighting. Slaughtering and plunder followed. The town and all the surrounding villages were burnt down."

Events such as these naturally did not conduce to amity between the two nations, and it was the state of affairs which resulted therefrom that first brought Great Britain and Russia on to the scene.

Russia, having doubtless in mind the tranquillity of her newly acquired provinces of Armenia and Georgia, wished if possible to avert a war; nor could Great Britain, with her millions of Indian subjects and her trade interests in the Gulf, afford to look quietly on while the two Mussulman neighbours went for each other's throats. It must be remembered that Persia was at that period a military Power, and possibly quite a match for Turkey, enfeebled as she then was by two recent wars.

At this time Sir Stratford Canning—"the Great Elchi," as he came to be called—was our Ambassador at Constantinople, and his unique influence with the Porte, combined with that of his Russian colleague, was successful in inducing the Turkish Government to substitute the pen for the sword. The Persian Government having consented too, a commission was convened at Erzeroum to discuss the frontier question. Great Britain was represented by three Commissioners—Colonel Williams (the famous defender of Kars in the Crimean War), Major Farrant, and the Hon. Robert Curzon,[1] a cousin of the ex-Viceroy of

[1] Author of 'Visits to Monasteries of the Levant.'

India. The last-named (whose literary genius admits of a fair comparison with that of the immortal Kinglake) has left in his 'Armenia' a vivid account of the extraordinary rigours and discomforts of the journey from Constantinople across the Black Sea to Trebizond, and from there on horseback over the mountains to Erzeroum, as well as of the rough, not to say dangerous, existence of the three Englishmen in that bleak Armenian town. He describes the primitive dwelling—half house, half stable—where the Commissioners were lodged in the company of a mixed society of nineteen lambs (who spent the day taking the air on the roof), one hen, a white Persian cat with her tail stained pink with henna, and a lemming "who passed his life in a brass-bound foot-tub." That any of the Commission survived to tell the tale is remarkable. The Persian representative was attacked by a furious mob of some thousands of Sunnis and besieged in his house for several hours; Curzon's two British colleagues were so nearly suffocated by charcoal fumes in their semi-subterranean home that they were only rescued in a fainting condition by the Russian Commissioner coming to pay a call; while Curzon himself had the following hair-breadth escape:—

"Mirza Jaffer, an old acquaintance of mine when he was Ambassador from Persia to the Porte, was too unwell to leave Tabriz, and Mirza Tekee was appointed Persian Plenipotentiary instead. On his arrival within sight of Erzeroum from Persia, all the great people, except the Pasha and the Commissioners, went out on horseback to meet him and accompany him on his entry into the town. There was a great concourse and a prodigious firing of guns at full gallop, which, as the guns are generally loaded with ball cartridge bought ready-made in the bazaar, though intended as an honour, is a somewhat dangerous display. Unable to resist so picturesque a sight, I had ridden out on the Persian road, though I did not join the escort, and, having returned, I was walking up and down on the roof of the house watching the crowds passing in the valley below, and looking at the great guns of the citadel which the soldiers were firing as a salute. They fired very well, in very good time, but I observed several petty officers and a number of men busily employed at one gun, the last to the left near the corner of the battery. At length this gun was loaded. A prodigious deal of peeping and pointing took place out of the embrasure, and, just as I was turning in my walk, bang went the cannon, and I was covered with dust from something which struck the ground in the yard in a line below my feet. On looking down to see what this could be, I saw a ball stuck in the earth; the soldiers had all disappeared from the ramparts of the citadel, and I found they had been *taking a shot at the British Commissioner.*"

But to return to our subject. The joint Commission sat at Erzeroum for four years, examining documents and interviewing important witnesses such as the Governor of Zohab and the Sheikh of Mohammerah, who travelled thither to give evidence. In 1847 their labours culminated in the Treaty of Erzeroum, by which the broad points at issue were settled, and a Commission of Delimitation was appointed to lay down the frontier on the spot.

This new Commission, on which Great Britain was represented by Colonel Williams alone, met at Bagdad in 1848, and proceeded to the southern end of the frontier. How evil a reputation for lawlessness the frontier then enjoyed is shown by the fact that their escort was settled at two battalions of infantry, a squadron of cavalry, and two guns, to be provided by each of the principal parties. Ostentation must, however, have been the chief motive when a gunboat with six guns was specially built to carry Dervish Pasha, the Ottoman Commissioner, from Bagdad to Mohammerah. Colonel Williams arrived at the latter place on the East India Company's *Nicrotis*, and after a great firing of salutes, as is reported, their work began.

From 1848 till 1852 they spent in travelling spasmodically up and down the frontier, their work delayed and disorganised by the obstructiveness of both the principals, and in particular by the erratic movements of Dervish Pasha—an over-zealous patriot, who thought to serve his country's cause by breaking away periodically on "tours of investigation," in the course of which he would erect a line of boundary pillars to suit his own personal views. At the same time the frontier zone was surveyed by British and Russian engineers, but all the efforts of Lord Palmerston and the Russian Foreign Minister could not reconcile the rival parties sufficiently to allow of a line being laid down on the ground.

The only records, so far as I know, of that journey exist in the Diary of General Tchirikoff, the Russian representative, which was published later by his own Government, and in a book devoted principally to archæological discoveries by Mr W. K. Loftus, an English archæologist attached to the Commission during part of its travels. Even Colonel Williams' official report is not extant, as that valuable record of four years' arduous toil, having reached England, had the misfortune to be dropped overboard near Gravesend, and found a sepulchre in the mud of the Thames. Negotiations continued after the Commission's return to Constantinople for rather more than a year, when the Crimean War broke out and brought them to an abrupt end.

When the war was over the frontier question was almost at once resumed. The first thing to do was to make a large-scale map of the frontier zone. For this purpose Commander Glascott, R.N., who had made the British survey, went to St

Petersburg, where he and the Russian surveyors worked at their respective maps till 1865. When at last the maps were ready they were—apparently for the first time—compared, the result being that by the time eight out of the seventeen sheets which composed each set had been examined, four thousand discrepancies in names, places, &c., were discovered. As it was clearly useless for the purpose in question to have two maps which were so very discordant, the surveyors returned to their drawing tables and, by some surprising feat of cartography, so manipulated the two versions as to produce a single copy known henceforth by the euphemistic title of the *Carte Identique.* This map, executed on a scale of one inch to a mile, was completed in 1869, just twenty years after the first surveys were begun,—its English co-author having in the course of his labours risen from the rank of lieutenant to that of post-captain in the Royal Navy—a unique record, one may reasonably suppose, in the annals of the science. The share of this country alone in the cost of production ran well into five figures.

Now that the representatives of the Mohammedan Powers could sit in academic comfort round a map and discuss things free from the petty vexations of Oriental travel, there seemed at last every hope of a satisfactory end to the dispute. But after five more years of negotiation and correspondence, the only progress made was an admission by Turkey and Persia that the frontier lay "somewhere within" the zone (a strip averaging twenty-five miles wide) represented on the map, and an undertaking not to erect any new buildings or otherwise prejudice each other's claims within that strip.[1] The explanation is perhaps to be found in those influences which Lord Curzon summed up as follows in his Romanes Lecture in 1907 on "Frontiers." "In Asia," he said, "there has always been a strong instinctive aversion to the acceptance of fixed boundaries, arising partly from the nomadic habits of the people, partly from the dislike of precise arrangements that is typical of the Oriental mind, but more still from the idea that in the vicissitudes of fortune more is to be expected from an unsettled than from a settled frontier."

Fresh frontier incidents were continually cropping up meanwhile. Anxious to avoid a rupture, England and Russia again intervened, and appointed delegates to assist in the discussion, the British delegate being General Sir Arnold

[1] This stipulation, though frequently broken in other ways, was adhered to *au pied de la lettre* in one rather humorous instance. It was proposed to link up the two countries by a line of telegraph wires along the Bagdad-Kermanshah route. Here, as almost everywhere, there was a strip of territory to which each side laid jealous claim, and a terrible difficulty arose. Turkish telegraph wires are carried on iron posts, Persian on wooden sticks (one can hardly grace them with a more imposing name). To set up a row of either across the debatable strip might prejudice the other side's claim. A solution of truly Oriental ingenuity was found to the problem—the line was put up with iron posts and wooden sticks *set alternately*.

Kemball. By a process familiar to any one who has watched the making of purchases in an Oriental bazaar, the two lines representing the rival claims on the map approached closer and closer by infinitesimal degrees. The *pazarluk* was almost concluded in 1877, when for the second time fate stepped in. The Servian War broke out, to be followed immediately after by the Russo-Turkish campaign.

After the Treaty of Berlin the frontier question was soon resumed, and once more a mixed Commission, including Sir Arnold Kemball, made the arduous journey to the frontier,—this time to investigate a local dispute concerning Kotur, a district not far from the new Russian boundary. Their efforts were entirely abortive, thanks to Turkish obstinacy, which has remained so persistent in the matter that this particular stretch of frontier (some thirty miles in length) had to be skipped over when the whole of the rest was finally demarcated in 1914.

By 1885 the British Government had spent, it is reckoned, over £100,000 in one way or another over the Turco-Persian frontier, with nothing to show for it except a few copies of a gaudily coloured map some twenty yards long and of doubtful accuracy, and the creation of a theoretical "frontier zone" which one of the parties at least made poor pretence of respecting.

There is nothing in particular to record besides a succession of unedifying disputes until 1906, when Persia's internal troubles became acute, and provided her neighbour with precisely one of those opportunities indicated in Lord Curzon's words already quoted. Turkey was quick to seize the chance, and before long her troops were twenty or thirty miles across the usually recognised frontier and deep into the Persian provinces of Azerbaijan and Kurdistan. These inroads continued for several years.

In the meantime the Anglo-Russian agreement was concluded, and Russia's sphere of influence in North Persia being now officially recognised, these violations of Persian territory became matters of concern to her as well as to the injured party. On one occasion a Turkish force actually came into collision with Russian garrison troops in Azerbaijan. Things were made worse by the waning of the power of the Persian Constitutional Government, and by 1913 so many regrettable incidents had occurred that yet again a Turco-Persian Commission met at Constantinople. Their sittings were so barren of results that again British and Russian mediation was called in. But this time the old farce was not to be allowed to repeat itself. Both the European Powers had now far more at stake: Russia because of her position in Azerbaijan, and England, among other reasons, because of the concession obtained by a British company for the monopoly of oil workings through a large part of Persia, some of the principal oil-fields discovered up till then being situated on debatable ground near the frontier zone. The energy of the respective Ambassadors at Constantinople had its result, and on November 4, 1913, a Protocol was signed by the Grand Vizier of Turkey and the Ambassadors of the three other Powers laying down summarily the frontier between the two Empires. A difficulty sprang up about the maps. Not one copy of the famous Carte Identique was forthcoming. The Turkish copy had, it appeared, been purloined by Izzet Pasha, the celebrated secretary of Sultan Abdul Hamid, for reasons best known to himself and his imperial master. All the other copies had disappeared. At length a battered tin cylinder, which for years had been accumulating dust in some corner of the British Legation at Teheran, was opened and revealed the searched-for map. The topographical difficulties were still too great, however, to allow of the frontier line being actually marked. At one place, for instance, the Turks resolutely claimed as their boundary a river whose name appeared on no map, and about whose position—or even existence—no one could give the slightest information. Farther south, for a distance of nearly 250 miles, the country was so uninhabited and little known that no data could be obtained on which to base even a general description of the frontier. The next step was to appoint a Commission of Delimitation, with the duty of settling the frontier line on the spot wherever it had been left vague and of demarcating the whole by putting up boundary pillars. British and Russian Commissioners were to participate in this, but under very different circumstances to those of the 1848 Commission. Their position—to borrow a metaphor from the "ring"—was changed from that of seconds to that of referees; in diplomatic parlance, they were attached as "arbitrating" instead of "mediating" members, and it was agreed that whenever the principals could not come to terms about any particular section of the line, they must refer the difference within forty-eight hours to the British and Russian members to arbitrate upon. The effect is illustrated strikingly enough by the fact that the 1913-1914 Commission finished the whole work, including a complete new survey of the frontier and the erection of two hundred and twenty-three pillars, in well under twelve months, as against the three years occupied by the peregrinations of their predecessors.

The Commission assembled at Mohammerah at the end of the year, and its experiences and adventures are set forth in the succeeding chapters.

The British Commission[1] (appointed partly by

[1] I must apologise for the loose use of the word "Commission" in the body of this book, meaning sometimes the whole Commission, sometimes the British section of it. The more correct term to describe the latter—"Delegation"—does not come easily to my pen, as "Commission" was the word always used, and I think the ambiguity, where there is any, is quite unimportant.

the Foreign Office, partly by the Government of India) consisted of the following officers: Commissioner, Mr A. C. Wratislaw, C.B., C.M.G.; Deputy Commissioner, Captain A. T. Wilson, C.M.G.; Chief of the Survey Party, Lieut.-Colonel C. H. D. Ryder, R.E., D.S.O., C.I.E.; Second in command, Major H. M. Cowie, R.E.; Officer in command of Escort, Captain A. H. Brooke, 18th (King George's Own) Lancers; Medical Officer, Captain H. W. Pierpoint, I.M.S.; Secretary, the present writer. Later on Captain F. L. Dyer, 93rd Burma Light Infantry, who was spending his leave learning Persian at Mohammerah when we arrived, was attached as Intelligence and Transport Officer.

Mr Wratislaw, a Consul-General in the Levant Consular Service, had some years before served at Basra as Consul and later at Tabriz as Consul-General, so was already well acquainted with the conditions in the extreme south and north of our route. Captain Wilson, who is an officer in the Indian Political Service, was formerly Consul at Mohammerah, and having travelled a great deal in the wilder parts of Persia, possessed invaluable experience of the Arab and Persian sections of the frontier. Both our Survey officers had acted on previous frontier Commissions on the borders of India and China, besides having accompanied Colonel Younghusband's expedition to Lhassa, the stories of which famous adventure always made our own journey seem depressingly tame and commonplace.

The Russian Commission, under Monsieur V. Minorski, corresponded to our own, but for the lack of a secretary and the addition of a naturalist, a most enterprising representative of science, whose enthusiasm remained undamped by the truly melancholy series of disasters which befell his collections. Major Aziz Bey of the Turkish General Staff represented Turkey; his deputy was also an army officer—in fact, there was only one civilian in their party. Persia's interests were upheld by Etela-ul-Mulk,[1] a member of the Teheran Foreign Office, and the non-military character of the Commission was compensated for by the addition of the Persian Director-General of Artillery, an up-to-date officer trained at the École Polytechnique and the Berlin Military Academy, who was attached as "Military Adviser." Each Commission had its own escort (the Russians having Cossacks), a doctor, and a staff of technical officers. The latter were all military engineers except the

[1] Every Persian, as I need hardly explain, who has attained to a certain position in civil or military life acquires an honorific and usually somewhat high-sounding title (which he often changes several times in a lifetime as he ascends in the scale). The Persian Commissioner's title may be rendered in English "The Brightness of the Empire." The Deputy Commissioner was "The Helper of the Sultanate," the Military Adviser "The Victorious Leader," and my fellow-scribe "The Beauty of the Kingdom"—an epithet which, I always imagined, must have been meant to refer to complexion of mind rather than of body.

Persian contingent, which consisted of an elderly and very devout gentleman of scientific tastes, who had been invested with the temporary rank of a General, and was fond of referring to himself and his colleagues as "we men of the sword" (though I do not think he ever girt one in his life), and two beardless youths, who were his pupils.

This is, very briefly, the story of a question which has worried Embassies and Foreign Offices for nearly three-quarters of a century. If its connection with our own country seems remote, it is the more striking as a specimen of the thorny bypaths along which the Government of a great empire inevitably becomes from time to time involved. We who have had, in the course of the last century or so, to delimitate between twenty and thirty thousand miles of our own frontiers, find ourselves forced by "the responsibility of empire" to lend a hand in settling other people's. Truly, to him who hath shall be given!

There is one noticeable characteristic of the Turco-Persian frontier which is due partially to the piecemeal and deliberate way in which it has been created. It supplies instances of practically every principle of delimitation known to the science. A frontier may be geographical, racial, linguistic, religious, or even purely artificial. That in question is all of these. In the broadest sense it is geographical, for it follows in the main a great mountain range. It is racial, in the south particularly, where it separates Iranian from Semitic, or, to be more specific, Lur from Arab; farther north, in Kurdistan, the division is, more correctly speaking, tribal. In the Pusht-i-Kuh region it is roughly speaking linguistic—smooth Persian dialect greeting your ear one side, and guttural Arabic on the other. Among the Kurds again the religious factor comes in, where an attempt has been made to leave Sunnis in Turkey and Shias in Persia. Finally, there is a stretch in the south where the frontier is frankly artificial, following stated lines of longitude and latitude.

Of physical features there is hardly one commonly used in frontier-making which we did not, at some time or other, apply: along the Shatt-el-Arab, "the line of low tide"; in the case of other rivers, one or other bank, or oftener still the "*medium filum aquae*"; in mountainous regions "the crest line" or "the watershed" (by no means synonymous terms), or else that much-disputed phrase "the foot of the hills." A certain section of the frontier was, as I have already said, fixed according to the "astronomical method" adopted so generally in Africa and North America, while at another point—where it was a question of creating an *enclave* round certain oil-springs—the line was drawn so as to describe an arc of circle. Strategic considerations came in here and there, as they do in the case of almost any mountain frontier, where the possession of a certain pass may be worth an army corps to either side.

To conclude this brief survey of the frontier, I must say a word or two as to how the Commission worked. About three-quarters of the frontier had been, as I explained before, delimitated at Constantinople, the remaining quarter being left blank. In the case of the former part the Ambassadors had, so to speak, sketched the main outline, leaving the Commission to fill in the detail. A section of frontier would be described in the Protocol as following the crest of a certain mountain, for instance, and passing thence to a neighbouring peak, leaving such and such a village to Turkey or Persia as the case might be. The line being thus broadly indicated, the engineers went ahead, surveyed a strip of country eight or ten miles wide, and had a detailed map waiting for the Commissioners by the time they arrived. The latter then, after examining the map and the ground, met in solemn conclave and debated the precise line of the frontier, which when agreed upon was described in writing, marked on the map with red ink, and on the ground with a line of boundary pillars. Along the undelimitated part the matter was, of course, less simple. The only basis for the Commissioners' guidance was the *status quo* frontier of 1848—a very illusive ghost, as one may well imagine. The rival claims in this region were often as much as twenty miles apart, and a compromise was next door to impossible, so that in the end almost the whole of this section was settled by Russo-British arbitration. A dilemma used to arise when the frontier ahead was quite undetermined, which reminded one of the ancient conundrum as to which came first, the hen or the egg. The problem was, how you were to settle a given stretch of frontier before you had travelled along it, and how you could travel along it if you did not know where it went.

The outward and visible signs of our labours were, of course, the boundary pillars. Their composition depended, inevitably, on the nature of the surrounding country. While we were in the desert, where the only available material was mud, we built thereof imposing pyramids, destined, no doubt, to provide the archæologists of some future age with subjects for excavation and learned dispute. In the mountains, if the spot was accessible for a mule loaded with mortar, a rough pillar of masonry marked the boundary line. To erect these we had taken with us a staff of *soi-disant* masons, and on the occasion of the first stone pillar we formed a respectful circle round these craftsmen while they plied their trade. Having collected a great quantity of rocks they marked a circle on the ground and set to work from within to build a wall—the only application of their art, I suppose, with which they were acquainted. Such was their enthusiasm, that before long they had completely immured themselves, and a hole had to be knocked in the side of their self-made tomb to let them out. The product of their technique was magnificent, but it was not a boundary pillar, so we decided to rely henceforth on the light of nature. Sometimes the approach to a site was so precipitous that no four-footed beast could be got up, and on these occasions a large cairn of boulders had to suffice. In places the piety of the inhabitants saved us two or three hours' hard labour when the summit of the mountain was already crowned by a *nazargah*, one of those heaps of stones so common throughout the country, which, to quote Fraser's 'Kurdistan,' mark "where some local saint performed a miracle, or, perhaps, ate his breakfast." As it is the duty of every passing Moslem to add a stone to the pile, the saint's memory and the boundary mark are alike safe from extinction. This, alas, is more than can be claimed for most of our pillars. No token of government is popular with the Kurds, and we should have known—even if the candid blackguards had not occasionally thought fit to tell us so—that our carefully built monuments were lucky if they stood for a day once the back of authority had been turned. Their demolition, however, did little harm to anything except our *amour propre*, as the precise bearings of each were, of course, carefully recorded and its position marked on the map.

There are, lastly, a good many side issues which a Boundary Commission has to deal with besides the actual demarcation of the frontier. To give examples, there were in our case anchorage rights in the Shatt-el-Arab, water rights in places where the frontier intersected streams or irrigation canals, and, most important of all, the pasturage rights of the wandering tribes whose annual migrations take them from side to side of the border. Such were some of the divers problems which occupied and sometimes perplexed the Commissioners, and whose solution depended in a great degree on the conciliatory genius of the British and Russian members—reinforced, when occasion required, by that most powerful solvent of human differences which hails from the shell-riddled slopes of Champagne. But having now introduced, perhaps at all too great a length, the Frontier itself, it is high time to begin the account of our acquaintance with it.

CHAPTER II.

MARSEILLES TO MOHAMMERAH.

I HAVE sometimes wondered what would be the result if you took the five hundred odd human beings who go to make up the passenger-list of a crowded P. & O., and boxed them up *on dry land* under just the same conditions as they live under on board ship. If no actual murder had been committed by the time you came to let them out at the end of a fortnight, at least I doubt if you would find a single one of the five hundred on speaking terms with the other four hundred and ninety-nine. There must be some magic about a sea-voyage which keeps most people, in spite of the monotony, on such excellent terms with themselves, each other, and all the world besides; but what the magic is I cannot guess, unless it be the pleasant anticipation of something awaiting them at the journey's end. In which case, our own prospect of a year or more under canvas and a thousand-mile march from the Persian Gulf to Mount Ararat should have assured us a particularly agreeable voyage when we sailed from Marseilles for Mohammerah on November 21, 1913.

The voyage, as things turned out, was extremely pleasant, but being "uneventful," as voyages usually are (which is to say, that we ran into nothing and no one fell overboard), I will not burden the reader with a detailed narrative of our three weeks at sea. In case he should, however, contemplate a similar trip to the Gulf at some future date, I venture to give a brief outline of the voyage with the help of a few extracts from my traveller's diary.

Ss. Mooltan, *Port Said, Nov.* 25.—Calm sea all the way from Marseilles. Flag-ship of Admiral Troubridge, Commanding the Mediterranean Squadron, in port here. We have taken on board the Christmas mails for Australia,—a record number of mail-bags, so they say.

Aden, Dec. 1.—Reached Aden last night and transhipped on to ss. *Salsette*, the Bombay "ferry-boat." Seeing the place for the first time at night, one's only impression is of a black mass of hill rising from the water's edge, with solitary lights twinkling on the slope and a row of bright illumination along the shore. One might be looking at the wooded slopes of St Cloud, with its brightly lit cafés lining the bank of the Seine. Reality, as seen by daylight this morning, very different. Nothing but the deadest, barrenest rocks and fort-like houses. We watched through the glasses strings of bizarre-looking, two-wheeled camel-carts, which seem to supply the local means of transport.

Ss. Dwarka, *Dec.* 5.—Left Bombay harbour this afternoon barely an hour after entering it on the P. & O. So much for my hopes of setting foot in India! It is fortunately only a thirty-hour run to Karachi, as the regular B. I. boat is laid up and the *Dwarka* is disgustingly small for what is often a rough passage.

Ss. Kola, *Dec.* 7.—Reached Karachi at midnight, changed steamers, and left again before sunrise. The skipper speaks cheeringly of the "Shimâl" we are sure to meet in the Gulf, and this boat turns out to be even smaller than our last. Really, this journey is like the "Rake's Progress"—we go one lower at each stage. Some of the other members of our Commission—Colonel Ryder and Major Cowie, our Engineer officers, with their Survey Party of about forty-five Indians, and Captain Pierpoint of the Indian Medical Service—were on board when we joined. Captain Pierpoint has a handsome setter-bitch and her puppy, who are to come with us on the Expedition.

Muskat, Dec. 8.—There is small-pox in the town, which prevents our landing. Nothing could be more picturesque, however, than the view from the bay. We approached past clusters of white pinnacle-rocks rising abruptly out of a deep blue sea, swung suddenly round a point, and came to anchor in a little natural harbour with rocky cliffs frowning down on it from both sides. At the farther end two rocks crowned by the ruins of Portuguese forts stand sentry on either side, and between them lies all you can see of Muskat—a sea-front of a dozen or so white houses built on a rocky base washed by the sea. There is a small coaling-station in one corner of the bay, and a fleet of high-prowed fishing-boats painted in stripes of white, yellow, and green is anchored in front of it. The face of the cliff on the opposite side is almost covered with the names, in huge letters of white paint, of all the British men-of-war which have patrolled the Gulf for generations past. Remains of Portuguese towers and bastions are visible all round, and we just get a glimpse between the mountains of the *hinterland*, where a force of troops from India is stationed at present to protect Muskat from tribal attacks, the interior being in a state of anarchy.

There is only one other ship in the harbour, an ex-tramp, which represents the Muskat navy, and is chiefly used, we are told, for taking the Sultan's wives on visits to Zanzibar, the two Sultans being cousins and on very friendly terms.

Dec. 9.—The promised "Shimâl" came. We were barely through the Straits of Ormuz and fairly into the Gulf when we ran into such a thick, black squall that the Captain turned the ship round for fear of running on to a reef. The worst of it was soon over and we were able to go ahead again. We have just stopped off a small island, a mere rock, where H.M.S. *Harold* is lying, to drop an engineer of Indian Marine whom we had on board. The *Harold* is typical of what our naval men have to put up with in the Gulf; she is nothing more than a Liverpool tug, with no accommodation but the deck, and spends her time cruising about these tropical seas after gun-runners. She carries two officers and a terrier-pup; the latter, not having seen one of his own species for months, went quite off his head at the sight of Sheila and Mick, the doctor's dogs, and tore madly up and down the top of the deck-awning barking himself hoarse.

Bushire, Dec. 10.—It seems that we are officially designated the "fast mail." This is not, as I at first supposed, a humorous allusion to our speed, which is a steady nine knots, but implies that our only ports-of-call in the Gulf are Muskat and Bushire. We are lying here in the roads four or five miles out from the town, and the cargo has to be lightered ashore. We put out several cases of whisky, which were loaded into special, so-called *haram* boats, as the ordinary lightermen are too good Moslems to have any truck with the forbidden liquid. The *haram* boats, it seems, are manned by crews of hardened agnostics, who, so far from refusing to handle the cases, are seldom known to deliver the same intact.

Talking to one of the ship's officers about cargoes for the Gulf, I have learnt two interesting facts,—one that there is a lively trade in Persian carpets from Manchester to Bushire (Persia's greatest port), and the other that the boats which visit Bahrein—where the great pearl-fisheries are — often carry consignments of sham pearls for judicious admixture with the real ones at the fountain - head, — at least so my informant assured me!

In the Shatt-el-Arab, Dec. 11.—We dropped Major Cowie at Bushire, whence he will exchange signals with Fao by cable in order to fix its latitude and longitude, and so give a starting-off point for the survey work of the Commission. Bushire itself is a "fixed point"—*i.e.*, it has been connected up with the survey system of India.

We reached the bar at the mouth of the Shatt-el-Arab early this morning, and, as usually happens, stuck. The most vigorous churning served only to cloud the sea all round with mud, so we had to wait till the tide rose and lifted the boat over. Just after entering the river's mouth we passed Fao on the right (Turkish) bank. Its only claim to notice is the Indo-European Telegraph Station, whence the cable crosses to Bushire on its way to India.[1]

The Shatt, of course, is a great, broad river, containing, as it does, the united waters of the Euphrates, the Tigris, and the Karûn between its banks; but the palm groves on either side cut off from view most of the country beyond, and what there is to see of it is mostly dismal-looking swamp. It is a lifeless scene, and the only moving things in it are the occasional flocks of wild-fowl

[1] Fao was the scene of the first engagement by the Persian Gulf Expeditionary Force in the present war.

which rise in front of our bows and wing their way up-stream, and a few naked riverain Arabs up to their waists in water cutting reeds, which they use for fish-traps. The banks are lined with these traps, which consist of narrow-mouthed enclosures so designed as to let the fish enter easily at high tide, but to give them no way of escape when the water falls again.

Some miles farther up—this time on the Persian bank—one rounds a corner and comes on a most incongruous sight. Rows upon rows of huge oil-reservoirs, of the same familiar shape as the great drums at a gas-works, stretch into the desert; along the bank in front stand bungalows, power-houses, and retorts for refining the oil, the whole linked together by a network of light railway along which little engines run clanking and shrieking. This turns out to be the Anglo-Persian Oil Company's refinery, the big enterprise which, like the Suez Canal, has the British Government as its principal shareholder. The oil-fields are 150 miles away, near Ahwaz on the river Karûn — from there the oil comes down in a pipe laid across the desert. In its refined state it is pumped into barges which lie alongside the bank—which, in turn, empty it into big tank-steamers anchored outside the bar.

Later.—We have arrived at Mohammerah, but have to put in several hours' quarantine before we are allowed on shore. It is a fine site for a town. The Karûn runs at an almost perfect right angle into the Shatt, and the town lies along its northern bank near the junction, the Union Jack over the British Consulate being nearly at the point. The Karûn here must be nearly as broad as the Thames at London, and is a fine fast-flowing river; its waters are bright yellow, and make tremendous whirlpools and eddies where they join the Shatt, so strong as to swing the bows of even an ocean-going steamer like our own—while it is a regular death-trap for small craft.

During the preliminary negotiations at Constantinople, Mohammerah had been fixed upon as the place of rendezvous of the four Commissions, and the date of meeting had been settled for the middle of December. We were the first to arrive, followed a few days later by the Russians, who travelled direct from Constantinople on board a Russian ship.

Our Turkish and Persian colleagues, however, entertained characteristically liberal views on the subject of dates and time-tables, and nearly a month passed before they appeared; though it must be conceded that the difficulties of their overland routes from Constantinople and Teheran respectively gave some excuse for the delay. It thus came about that we spent in all six weeks in camp at Mohammerah, and had ample time to fit out the expedition and see something of the neighbourhood.

Before it emerged from its obscurity into the

grim lime-light of war, the country at the head of the Persian Gulf was, I imagine, to the average person little better than *terra incognita*—a region only vaguely apprehended in connection with dates, oil-fuel for the Navy, or the Garden of Eden. Possibly I wrong the country—or the "average person"—but the vivid recollection of my own furtive and shamefaced hunt for an atlas when I first got orders to go to Mohammerah, encourages me to enter on a brief description at the risk of repeating well-known facts.

In the first place, all Mesopotamia is Arab country. The neat line which provides a northern limit to Arabia on many of our maps has no basis in fact, and even the frontier between the Turkish and Persian Empires, in this particular section, marks no ethnical or linguistic boundary. Indeed, the proportion of racial Turks in Irak and of racial Persians in Arabistan is almost negligible, and the spoken language is Arabic throughout.

Compared with the real Bedouin of the Arabian desert, the Arabs in the parts I am speaking of are a degenerate lot; they are largely agriculturists, and many of them live in towns like Basra and Mohammerah. They are divided into several large tribes—the Muntefik, the Shaab, the Beni Lam, Beni Turuf, and so on—and, with the exception of some of the riverain Arabs, are more or less nomadic, often sowing a patch of land with some crop or other and leaving it to grow till they return again after some months to reap it. The same spirit is reflected in their domestic architecture, which takes the form of reed and palm-leaf huts, which can be put up in a few hours and abandoned without regret.

These Arabs have a good many unamiable traits, not the least of which is their love of piracy. As long ago as 1780 (the letter exists in the archives of the Bushire Residency) the Factor of the East India Company at Basra wrote to his superiors at Bombay praying them to despatch a sloop to the Shatt-el-Arab to protect his cargoes lying in the river. The precaution is necessary even now, and previous to the war a gunboat was sent annually up the river at the season when the dates are being shipped, to protect the boats from pirates. These river vermin practise their trade in its most dastardly form, creeping on board ships at night and murdering the crew. Even during our short stay a British captain at Basra sleeping on board his vessel was killed in this atrocious way, the villains escaping arrest.

This immunity which the pirates generally enjoy is due to the ease with which after an exploit in Turkish waters they can escape to the Persian shore, or *vice versa.* Some years ago the British India mail-boat ran aground and a gang of pirates boarded and gutted her. They escaped, as usual, on to Persian territory, but this time, thanks to the energetic help of the Sheikh of Mohammerah, they were caught and brought to book and a large part of the cargo recovered. As a graceful tribute to commemorate the event, the B.I. steamers always fired a one-gun salute as they pass the Sheikh's palace on their way up the river. The salute was returned from the Sheikh's battery on shore—not without some risk to the unfortunate gunner who used occasionally, when a certain young Third Officer fired the ship's gun, to get a charge of deck cricket balls whizzing round his ears.

The main features of the country can be summed up in three words—river, desert, and marsh, the river being, of course, the essential feature. The "Waters of Babylon," which once made Mesopotamia a rival with Egypt for the title of the "World's granary," still keep their fertilising powers intact. But the old dams, canals, and barrages are gone, and the productive land is now narrowed down to strips of palm grove fringing the river banks. Where the palm groves end the desert abruptly begins. There is nowhere that

> "Strip of herbage strown,
> That just divides the desert from the sown,"

where old Khayyám invites us to wander in blissful oblivion.

The desert itself is not of the good honest sand that one usually associates with the name. In the south at least, where it was once part of the delta, it is rich alluvial soil, good going for man and beast when dry but turning to the veriest quagmire of mud after rain or floods. The marshes have a character of their own, but I will speak of them again later.

The site of the Garden of Eden is located at Kurna, where the Euphrates and Tigris join—or so the school of pre-Darwinian theologists would have us believe. The crab-apples which grow to-day by the river's edge do indeed lend colour to the theory, but—unless conditions have radically changed—there is one, I fear, insuperable objection. Our erring ancestor's chief lament when ejected from Paradise was that he had to turn to and work. Now, in this country, nobody works—at least not what you can call work. Nature, by a bountiful provision, which says little for the good dame's powers of discrimination, absolves the inhabitants from all such necessity. The high tides at the head of the Gulf so bank up the Shatt as far as the junction of the rivers, that all the lazy Arab has to do is to cut a few canals and ditches and let the rise and fall water his land automatically twice a day. The date-palms, too, demand a minimum of labour, and in return for it provide him generously with food, fuel, and building material. I should doubt if there is another place in the world where the people do so little for their living, except perhaps the South Sea Islands.

Dates are exported to Europe in enormous quantities every autumn, and the export trade is chiefly in the hands of British firms established at Basra. Basra itself is not actually on the river, but lies about two miles inland, in a big clearing among the palms, and is a rather mean and dilapidated town—unless eighteen months of British administration has instilled into it a greater degree of self-respect. A fair-sized creek connects it with the Ashar, which is the port for the smaller vessels, but the big ships lie anchored in mid-stream opposite a long line of substantial brick buildings where the European merchants live and do their business.

It was from Basra that Sinbad the Sailor used to start on his voyages of adventure. At the time when his fabulous journeys were supposed to have taken place, Bagdad was at the height of its glory. The court of the Caliph, Haroun-er-Rashid, attracted thither all the riches and wits of the Mohommedan East, and the Caliph lived in a state of lavish splendour which has seldom been surpassed. One can dimly picture the scene at the port of Basra when a fleet of *dhows* sailed up from the Gulf laden with the slaves and pearls and spices which honest Sinbad describes with such relish. In these prosaic days, alas! you find instead a line of rusty sea-tramps full of Manchester cottons and iron-ware lying in the anchorage. At the Ashar itself, however, there are plenty of *dhows* to be seen, with their graceful, tapering masts and towering poops, painted often with gay designs in all the colours of the rainbow.

Thanks to the war, Basra has become a familiar name in our newspapers, while Mohammerah, lying on the other shore, has received at most a passing mention. But our connections with the Persian town have grown so much of late, and its progress promises to interest us so closely in the future, that, if the reader's patience can hold out a little longer, I should like to give one or two historical and geographical facts about it. Mohammerah also has been entered and occupied by a force of British troops—but some time ago; to be precise, in 1856. The occasion was the Persian war. That little-sung campaign fills very few pages in our military annals, and was remarkable chiefly for the almost complete lack of resistance on the enemy's part and the very short time it lasted. Persia had attempted to seize Herat, which we countered by despatching a force up the Gulf, bombarding Mohammerah, and sending a regiment—the Seaforth Highlanders—up the Karûn to capture the town of Ahwaz. The whole affair only lasted a little over four months, and now a few bits of shell ploughed up from time to time on the outskirts of the town and some gashes in the trunks of the older palm-trees are almost the only records of it left.

The importance of the place from our point of view is partly due to the creation of the Oil Company, who have their headquarters and a large English staff at Mohammerah, and partly to its position at the junction of two large navigable rivers. Persia is so cut off by mountain-barriers from its neighbours that the trade-routes to the interior are exceedingly limited. There are, of course, the Gulf ports, such as Bushire and Bunder Abbas, feeding parts of Central and Southern Persia, and in the north there are the Caspian ports and the Russian railway to Julfa, but throughout the 1200 miles of land frontier which separate the borders of Russia from the Gulf only two routes of any importance exist. One is the road from Bagdad to Kermanshah, which is closed as often as not by the turbulence and brigandage of the tribes; the other is from Mohammerah up the Karûn. Merchandise is transhipped at Mohammerah from the ocean-going steamers into antiquated stern-wheelers, carried up as far as Ahwaz, and thence distributed; but a scheme was on foot before the war for constructing a railway from Ahwaz to Khurremebad, which would enormously increase the importance of the route. A preliminary survey was made, and a party of British engineers went out to the spot in 1913; the rapacity of the tribesmen on whose goodwill the work depended was so bottomless, however, that the scheme was still held up when the war began.

The Sheikh of Mohammerah, Sheikh Khazal, K.C.S.I., K.C.I.E., has always been a loyal friend of the British, and his friendship has now stood the severe test of the war in Mesopotamia and the revolt of a large part of his subjects. He is a middle-aged Arab of liberal tendencies, though far too cautious to introduce any wholesale, ready-made projects of "civilisation" within his domain. Progress has to be a gradual affair in Mohammerah.

His authority extends over the two tribes of the Shaab and the Muhaisin, which have to a great extent coalesced under their common ruler. These tribes are supposed to have immigrated from Arabia some 250 years ago and settled in Persian territory. Thanks to their remoteness from the centre of government, their difference of language and race, and the incessant rivalry between Turkey and Persia, they have remained very independent of the Shah and his Government, and the present Sheikh, having steered adroitly through the troublous waters of the Revolution, is now sovereign in all but name over the greater part of the province of Arabistan, a country nearly as big as Belgium. He levies his own taxes and makes his own laws, and the Imperial Government has only a shadowy representative in the person of the *Karguzar*, or Foreign Office Agent. The Sheikh's prerogative does not, however, extend to the Customs, though even here he is *ex officio* Director-General.

One other token of the Shah's authority I had all but forgotten — the Persian Navy. It lies anchored in perpetuity off Mohammerah, and consists of two pre-Dreadnoughts, once merchantmen, commanded by an ex-captain of the Belgian Mercantile Marine. Rumour has it that the Navy is immobilised owing to the fact that the flagship is without her engines, these necessary adjuncts to locomotion having been privily sold by a former commander, if rumour speaks true, when that gallant officer was particularly hard pressed by his creditors. However that may be, the vessels are kept outwardly smart and in good trim, and present a fine spectacle when, rigged out with bunting, they fire a royal salute on the Shah's birthday.

CHAPTER III.

PRELIMINARIES TO THE START.

CHRISTMAS passed at Mohammerah with such a round of festivity and good cheer as to obliterate any feeling of exile on a far-off shore. Hospitality is practised in the Gulf in that open-hearted fashion which is the characteristic of English people in out-of-the-way parts, and our Christmas dinner at the Consulate with Major and Mrs Haworth, and the New Year fancy-dress ball at the house of one of the oldest British residents at Basra, were memories to be cherished fondly in the lean days to come. The mention of good cheer, by the way, brings to mind a little anecdote which is told of a wealthy and influential native merchant of Mohammerah, and is a nice illustration of the universality of that painful obligation which falls at times on the best of us, namely, of "keeping up appearances."

Hajji Ahmad, the hero of the story, is a personage much respected for his wealth and piety, but he has, I grieve to say, acquired from his English friends such a taste for whisky that he finds it hard to get through the morning without his "tot." A friend of his died one day, and after attending the funeral he joined a gathering of pious folk collected for the performance of religious exercises in memory of the deceased. Each man, as the custom is, had brought with him a little teapot full of water to moisten his lips when parched with chanting the Koran. Our Hajji's teapot contained a less innocent beverage, the odour of which reached the nostrils of a holy Mollah seated opposite. The Mollah, scandalised beyond words, rose and denounced the Hajji to the assembly. A horrified murmur ran round the room, but the Hajji's wit did not desert him; turning to a respectable fellow-merchant at his side with whom he had an important contract in prospect, he handed him his teapot, and with a surreptitious nudge of the elbow, bade him drink. The merchant took a gulp, and barely concealed the agony of his burnt gullet. "Whisky or water?" asked the Hajji in uncompromising tones. "Water, by God," loyally replied the merchant, realising the issue at stake. Whereupon the Hajji rose, full of righteous indignation, and hugging the teapot carefully to his bosom, marched from the room, vowing that he would have nothing more to do with such infamous slanderers.

Whence it is clear that there are religious hypocrites all the world over, who share the old pharisaical view concerning the "inside of the cup"—or, as in this case, of the teapot!

The streets of Mohammerah are filthy beyond the power of description, but as all the European bungalows, including the bank and the club, lie along the river-bank, one goes everywhere by water, as if one were in Venice, and so avoids their horrors. For this purpose there exists a particularly well-adapted form of craft called a *belem*, as characteristic and almost as fascinating a boat as the fast-disappearing *caïque* of Constantinople. In shape it is a hybrid between a canoe and a punt, while it has a slight *soupçon* of the gondola added; that is to say, it has the general lines of the first, the flat bottom of the second, with a graceful curl of the bow and stern reminiscent of the third. The passenger reposes in cushioned ease in the middle of the boat, with two rowers in front of him and one behind. I call them "rowers," but in fact their functions are no less hybrid than the craft. When going against the tide they pole along near into the bank, using their long flexible bamboo poles so cleverly that the progress is quite surprisingly rapid and smooth; if, on the other hand, the tide is favourable, they unship round-bladed paddles, and your *belem* glides in mid-stream with an occasional quiet stroke to hold it on its course.

The town itself straggles for nearly a mile along the northern bank. Like many other Eastern cities it possesses a meretricious charm, hiding away an ignoble jumble of mud-built houses behind a singularly picturesque river front. Sliding past it in a *belem*, the impression one receives is of blue-painted balconies, tented coffee-shops, and fragrant gardens, with a few intervening creeks spanned by high-arched bridges of mellow brick-work.

While we were still busy with the thousand and one preparations for our journey, buying horses, packing kit, engaging servants, and so forth, the last of our belated colleagues arrived, and the official frock-coat *visites de cérémonie* having been duly exchanged, the work of the Commission began. The first meeting was held at the Prime Minister's house; he is a rich Persian merchant who, at the mature age of sixty, had left his native land some months before to pay his first visit to Europe and "Parees." His son, however, did the honours in a fine reception-room, hung round with forty or fifty oleograph portraits of crowned heads and presidents, past and present, ranging from the Shah of Persia to Kruger. The Turkish Commissioners differed from the rest in being soldiers instead of diplomats; indeed their whole party was organised on strictly military lines, very much to the disgust of my "opposite number," to use a naval phrase, who, being a Foreign Office Secretary accustomed to a civilian's life of ease and luxury, did not at all appreciate the rigours of field-service conditions.

Our first piece of active work was a trip down the Shatt to the river's mouth, on board the gunboat *Marmaris*, the only war vessel, besides a couple of armed launches, which the Turks kept in these waters. The frontier follows the left bank of the Shatt, so all that had to be done was to fix the point where it first strikes the shore from seawards, and settle the case of a few small islands near the Persian bank. A few days later the solemn act was performed of erecting boundary pillar Number 1.

The occasion was one of great ceremony and not a little humour, and deserves, I think, some description. The site of the pillar is at Diaiji, a village a few miles up-stream from Mohammerah, where the frontier leaves the river to strike northwards into the desert. The whole Commission took part, and embarking this time on a Persian steamer (to equalise matters), steamed up to Failiyeh to pick up the Sheikh and his eldest son.

The palace faces on to the river, and is chiefly remarkable for an alarming wood-work erection painted brightest blue, supporting two heraldic shields, emblazoned with the Sheikh's initials in European characters. Adjoining the palace is his Highness's private theatre, and round about clusters a considerable village inhabited by his bodyguard, a ragamuffin crew of armed men, mostly of negro blood, who accompany their master wherever he goes—a necessary precaution in a country where the assassin's knife has accounted for more than one ruler.

We had towed up a small flotilla of boats, into

which we embarked at the mouth of a canal. The Sheikh, who wore a long red cloak, travelled in a *belem* double the ordinary size, while the bodyguard kept pace along the bank, not in the least embarrassed by having to wade waist-deep across side canals every few minutes. At one point there was a village on the bank of the canal, with a high wall outside. As the Sheikh's boat drew level with it, a chorus of feminine voices from behind the wall raised a chant of "Hosanna," greeting their chief as the people of Jerusalem used to greet their king. We proceeded up the canal for two miles in a stately procession, till, the tide falling, we found ourselves aground on the mud, and had to take to our legs. The Sheikh, after being shoved—*belem* and all—for some distance over the mud by the united strength of the bodyguard, and its being beneath the dignity of a Sheikh to walk on foot, had perforce to return home, leaving his son to accompany us and do the honours. The latter is a handsome youth, sufficiently civilised to drive a motor-car, and at the same time oriental enough to have been recently relieved of his governorship of a town for having too many merchants beaten to death!

Our path led through palm groves, and a very sporting element was introduced by the palm-trunk bridges, over which we had to cross the innumerable small canals. Formed of a single trunk sagging heavily in the centre, these bridges provide one with all the sensations of a tight-rope dance, and once on them it was difficult for some of our more corpulent colleagues, arrayed in their smartest uniforms and decorations, to retain the dignity they possessed on *terra firma*. After one or two partial immersions we arrived at the site of the pillar, and began the work of the day. First came the gruesome sacrifice of a poor sheep, whose life not even the British Commissioner's prayers could save. A stake was then driven in, and each of the four Commissioners having laid a brick round its foot, a body of local masons completed the edifice. When I spoke of the "work of the day" I was mistaken; that began after the pillar had been built. We walked back to the village, and there found a huge marquee erected by the Sheikh for our reception. We entered—and were dumb. Imagine a huge tent with tablecloths laid on the floor round three sides, and on them set such a sea of dishes that any attempt to count them was hopeless. I made a rough calculation by counting the dishes on a single section, and multiplying by the number of sections; the result worked out at something over two hundred and fifty. Four whole roast sheep formed the *pièces de résistance;* around them were ten dishes, some eight feet in circumference, heaped mountain-high with *pillau,* and each crowned with a roast lamb, twenty or thirty fowls on smaller dishes, and innumerable bowls of rice, hashes, and junkets. We sat reverently down before this unheard-of profusion on a row of beautiful carpets. A touch of the vulgar West was, alas! introduced by the drinks, which consisted of whisky and porter; however, the latter seemed in keeping with the roast sheep, so we elected to drink that. All trace of the Occident was entirely dispelled, however, by the arrival of a grizzled old nigger in a long linen dress, and a beltful of cartridges round his waist, who acted the part of butler. The dishes in the centre of the cloth were, of course, far out of one's reach, so the old fellow kicked off his shoes, trussed up his skirts, and stalked boldly on to the "table." He wielded an enormous ladle, with which he piled up your plate from whichever dish you chose, unless he thought you looked particularly hungry, when he took your plate, and burying it bodily in a mountain of *pillau*, handed it back heaped a foot high. Finally, he tackled a whole sheep by the very simple process of seizing the body with one hand and a leg with the other, giving a hearty wrench and—handing you your joint. I was sitting fascinated by his huge black feet wandering unscathed among the multitude of dishes, when he turned round abruptly to give me my plate, lost his balance, and put his foot splosh into a luscious dish of apricots and chicken hash. He was not a whit disconcerted, however, but passed serenely on his way, leaving behind him an intricate design in brown and yellow on the tablecloth. We finished the banquet without making the smallest visible effect on the piles of victuals, and retired to coffee and cigarettes, while the retainers were let loose on the feast; half an hour later there was not a grain of rice left! In the meantime the tide had risen, and to every one's intense relief the *belems* were found waiting only a few hundred yards away to take us back to the ship. What would have happened if we had had to face again the single palm-trunk bridges, I shudder to think!

Our escort arrived from India a few days after the rest of us. It was commanded by Captain Brooke of the 18th (King George's Own) Lancers, and consisted of a native officer and 30 *sowars*. Captain Dyer of the 93rd Burma Light Infantry joined as Transport Officer soon after, and so our numbers were complete. The entire party, including 8 English officers, the escort, an Indian clerk, 4 Indian surveyors, and 40 survey-*kholassis*, servants, syces, *farrashes*, muleteers, and camp-followers of every description, totalled about 150 men. The Russians and Persians numbered rather fewer, while the Turks, who travelled very light, were scarcely more than 50. The Russians, I must remark *en passant*, emphasised the cosmopolitan nature of the Commission by bringing with them a Chinese *dhobi*, picked up heaven-knows-where, but a first-rate performer with the soap-suds!

The very formidable task of arranging beforehand for the transport and provisioning of such a party as our own across nearly 400 miles of almost uninhabited desert had fallen on the broad shoulders of the Deputy Commissioner, Captain Wilson, who had preceded the rest to Mohammerah for this purpose. He knew the country well, having acted as Consul at Mohammerah and travelled largely in Arabistan, as well as in the wilder regions of Luristan, which lies beyond the mountains; but armed as he was with an intimate knowledge of local conditions, a wide-spread reputation among the Arabs, and a quite inexhaustible fund of energy, it was still no light matter.

No mules were obtainable locally, so the whole complement of 230 odd had to be hired from places such as Shuster and Hamadan, the latter as much as 300 miles away. Camels were to be had, but an initial experiment with 50 of these beasts proved them, even in their native element, less suitable than mules for the sort of work in hand.

The latter appeared one morning out of the blank desert behind our camp, heralded by a great rattle of hoofs and carillon of bells, and accompanied by smiling muleteers apparently as little concerned as the beasts themselves at the prospect of a journey to Ararat and back. They formed such an intrinsic part of our existence throughout the expedition that I will start by giving the best idea I can of what a Persian mule-train is like. The owner accompanies his own beasts; if he is the proud possessor of 20 or more he is usually mounted on an arab mare, leaving *yetims* (*anglicè*, orphans) actually to drive the teams; if he has only half a dozen or so he goes on foot, with a diminutive donkey to carry his belongings and himself when tired. Every bunch of fourteen or fifteen mules is led by a *yabu*, a pony who performs the duty of a bell-wether to a flock of sheep. The mule carries his load on a wooden pack-saddle, a cumbersome high-peaked thing which never leaves his back day or night while on the march. The poor beast is thus debarred from that greatest of mulish joys—rolling, and it is a truly pitiful sight to see him, released from his load, lie down on the sand and struggle wildly to get his hoofs in the air, his efforts perpetually baffled by the projecting peak of his saddle.

The musical lady of Banbury Cross is quite thrown into the shade by the Persian pack mule, and even more so by his companion, the *yabu*. Though deterred by nature from wearing bells on his "fingers and toes," he more than makes up for it by the quantity which hang from every other part of his anatomy and transform him into a kind of ambulant belfry. An inventory of the trappings of a really self-respecting *yabu* would be somewhat as follows: Beginning at his head, he wears a headstall hung all over with tassels and little round bells, and thickly encrusted with blue-and-white beads edged with cowrie shells; the headstall broadens out over his forehead where it frames a round piece of looking-glass, giving him a rather cyclopean air, while between his ears there nods an imitation bird the size of a thrush, covered also with beads. His neck is encircled by a broad strap handsomely encrusted in the same manner as the headstall, from which dangles a set of bells of ever decreasing size, fitting within each other after the fashion of Chinese boxes. His gaudily-coloured breeching is edged with scores of little bells, while to crown all there hang suspended on either side of his saddle, so as to almost brush the ground, two colossal bells nearly two feet high. I had the doubtful privilege of having a particularly swell *yabu* in my team whose owner had the distressing habit, on arriving in camp after it mattered not how many hours hard marching, of urging his beast into a furious gallop, while he executed a kind of devastating musical ride in and out of boxes and tent ropes, with such a chiming and clanging of bells as never was heard, and to the serious detriment of the contents of my *yakdans* bumping and bouncing on the poor animal's back.

The *charvadar*, as the Persian muleteer is called, is usually a cheerful fellow in spite of his hard life, and ours were no exception. When not on the march they spent all their spare time sitting beneath their low black shelters—they are too shapeless to be called tents—sipping eternal glasses of tea, gossiping and smoking their long-stemmed pipes, which when not in use they carry stuck through their cummerbunds; but during the idle days at Mohammerah they emulated the more active pastimes of the Indian *sowars*. One evening, hearing a great hullabaloo from behind the camp, we went out to see the cause, and found a tug-of-war going on between a team of *sowars* and another of Persian muleteers. The Persians were lusty fellows and were putting up a very good pull, encouraging each other all the time with shrieks of "Allah." Some bare-legged shepherds driving their flocks home had stopped on their way to watch the fun, and they in turn were presently joined by a party of the Sheikh's henchmen, Biblical-looking figures in the traditional flowing Arab dress, carrying long-barrelled, silver-bound rifles. Seeing after a while that their countrymen were in danger of being hauled over the line, the onlookers could contain themselves no longer, but rushed madly in and seizing the rope brought the proceedings to an abrupt close. Afterwards a party of our Persian servants and hangers-on indulged in the ancient and—it would seem—universal game of leap-frog. They were a queer crew. The "frog" was a genteel-looking individual, in a long blue frock-coat, with the Royal Arms on his hat, and was leapt over by a succession of the wildest creatures in ragged red or brown shirts, and bell-mouthed trousers reaching half-way down their legs, with greasy black curls waving, as they ran, from under the black felt *kullahs*—a form of head-gear which I can only compare to a French *casserole* turned upside down.

The sporting proclivities of the *charvadars* found their fullest scope, however, on the day of our gymkhana. It was a very *pukka* gymkhana. There were horse-races, foot-races, sack-races, tent-pegging, trick-riding, and a full-dress mounted display by the Lancers, but *the* event of the day was the half-mile mule race (owners up). There was a field of fifty, and they duly ranged up at the end of the course. But whereas one man can get a mule to the starting-point, no power on earth can make him start in any direction but the one he chooses. The direction these particular mules chose was at right angles to the racecourse, and in two minutes they had disappeared among the date-groves. They were duly retrieved and again faced the starter, and a minute later the whole lot came bucketing down the course with the thunder of a cavalry charge, led by our Deputy Commissioner mounted on a big white horse which performed the function of the humble *yabu.* Fifty mules with bells clanging and clashing, mounted by fifty madly excited riders clad in every colour of the rainbow, and bouncing like peas in the huge wooden saddles, galloping pell-mell in the wake of a British officer cantering along with all the dignity of the parade-ground, was a sight for the gods. The *charvadars* wondered why we laughed!

The subsequent prize-giving was perhaps the most picturesque part of the show, when Mrs Haworth, the wife of the Consul, graciously presented prizes in turn to gorgeously-uniformed *sowars*, tattered camp-followers, respectable English residents, and, last but not least, to the hero of the mule race, a sturdy but bashful little fellow barely 5 feet high, with a long henna-stained beard, a green shawl round his waist, and gnarled brown legs bare from the knee downwards.

The work of the *corps technique* had all this time been going on apace, most of it devolving on the British members. It did not proceed without a slight mishap. Fao having been successfully fixed in relation to Bushire, Major Cowie set off thither one windy day to carry the connection up to Mohammerah. He took a motor boat and put his servant, kit, and four days' provisions in a *belem* which they towed behind. The *shimâl* was fierce and the river rough, and when half-way to Fao the *belem* shipped a wave and sank. The servant was rescued by the scruff of his neck, but kit and provisions went to the bottom, where they doubtless found a billet in the maw of a shark. The motor boat breaking down soon after, the unfortunate Major had to live on such charity as Fao can provide till he caught the next up-mail four days later. In the meantime, however, a British naval record was discovered giving the exact position of Mohammerah, and a base having been accurately measured at Mohammerah itself, a traverse was run up the Karûn to Nasiri near Ahwaz. It was from there that the regular triangulation was begun, on which is based the new map of the frontier.

CHAPTER IV.

THROUGH THE LAND OF ELAM (1).

The winter rains were comfortably over when, on February 14, we left Mohammerah on the first stage of our journey. The surface was dry but soft after the frequent floods of the last two months, and made the going perfect, while the air had that matchless combination of warmth and crispness which you find only among mountains or in the desert in early spring.

But before telling of the journey I must sketch roughly its scheme. The frontier, after leaving the celebrated pillar No. 1 at Diaiji, runs for sixty or seventy miles due north, turns due west for another twenty, and finally northwards again till it reaches a spot called Umm Chir, or "The Mother of Pitch." Beyond that there is a great stretch of 200 miles, where it goes in a fairly straight line north-west to Mendeli, situated on a latitude slightly north of Bagdad. These details are necessary to explain the Commission's wanderings. As far as Umm Chir the frontier could be marked on the map but not on the ground; for the reason that the first part of it runs through an arid desert too dry for travellers to pass through, the second part through an immense marsh (the Khor-el-Azem) which is too wet. The desert and the greater portion of the marsh being uninhabited, there was, moreover, no need for pillars even if it had been possible to erect them; so the frontier was made to follow convenient lines of longitude and latitude and left to look after itself. A rendezvous was fixed at Umm Chir, giving each Commission a fortnight to get there by whatever way it chose. The route selected by the British Commission formed roughly two arms of a triangle —the first arm running through the Arabistan desert parallel to the Karûn for about sixty miles; the other set at an obtuse angle towards the north-west, and passing along part of the river Kerkha and the northern edge of the great marsh.

One last thin thread of civilisation accompanied us for the first day or two of our march into the wilderness—to wit, the Persian State Telegraph. The line had been almost always reported "out of order" when we had wanted to send telegrams from Mohammerah during the previous six weeks, —once seen, what amazed one was that it is ever *in* order. A row of inebriated wiggly posts stretches across the desert; some have staples driven in to carry the wire, some have not. In the latter case the wire is simply twined round the post. Where two lengths join, the ends are twisted casually together, and sometimes for as much as a hundred yards on end the whole thing trails along the ground. And yet messages have been known to get through!

The desert at this season of the year, where not too much impregnated with salt, is covered with patches of young grass as smooth and as fine as the lawn of a cathedral close. Among the grass grow tiny aromatic plants, almost indistinguishable to the eye, but filling the whole air with a pleasant, keen smell. The effects of mirage are often startling. Our caravan, when on the march, straggled over two or three miles of country, and to any one riding somewhere near the middle the head and tail of the procession seemed always to be marching through a smooth, shallow lake; occasionally, for some unfathomable cause, the mules and men would execute a bewildering feat of "levitation" and continue their progress in the sky. Often we saw a lake spread out on the horizon, stretching a long arm towards us to within a few hundred yards; at other times a clump of palms or a group of mounted men appeared in the distance, only to resolve themselves, as we approached nearer, into bushes of low desert scrub or a grazing flock of goats.[1]

[1] Speaking of mirage, a curious incident is reported to have happened during the fighting between Fao and Basra at the beginning of the Mesopotamian campaign.

Our men, after a particularly courageous attack across the open desert (which at the time was such a sea of mud that they had to advance at the walk), reached the Turkish trenches and put the

Sleep was hard to woo on our first night in the desert. The muleteers, either through laziness or fear of thieves, always left the bells on their mules and picketed the animals in line outside our tents. It takes some time to attune one's ears to the unwonted music, and what little sleep this "nocturne" spared us I and my tent-fellow sacrificed through ignorance of Persian custom. Both our Persian servants sported ostentatiously large European watches. We gave the order to be called at six *alla franca*. The noisy eruption of my boy at half-past three woke us from our recent slumbers. Curses and missiles having convinced him of his error, he fled—only to be followed by my companion's boy, with the same ill-placed zeal, an hour later. The watches, had we known it, were but a token of gentility, and implied no ability on their owners' parts to read the dial. In Persia you must learn to use a looser phraseology in regard to time, and regulate your hour of being called by "the break of dawn" or "sunrise."

On the afternoon of the second day's march

Turks to flight. The enemy were now in the same predicament as the British had been in just before, and provided a splendid target for our artillery as they floundered through the mire. A gunboat was lying in the river, and the men in the tops were watching the proceedings when they were surprised to see our guns suddenly stop firing, although the Turks were still easily within range. It transpired later that, to the eyes of the gunners on the desert level, the target had *disappeared into the mirage*.

from Mohammerah there comes into sight a solitary group of trees, a pretty sure sign in this denuded country of the sacredness of the spot on which they grow. There among them, truly enough, gleamed the white dome of a saint's tomb. On reaching the place, we were puzzled to find the ground all around strewn with a number of shapeless forms covered with reed-matting. Wilson, knowing the country well, supplied the explanation. The buried *seyyid*, it seems, stretches an arm of protection over all objects within a certain radius of his tomb, and such is the odour of his sanctity that not the hardiest thief dare touch a thing within the circle. The shapes lying round were ploughs, hoes, and any other bits of property belonging to the semi-nomadic country folk, who, when the season's work in the fields is over, leave them here till they return in the following spring. The place has thus become a kind of "safe-deposit" for the entire neighbourhood.

But another, somewhat gruesome, class of goods is warehoused round the tomb—to wit, corpses. The warehousing of a corpse sounds a trifle indecent, but it is literally what takes place. Every true Shia cherishes in his heart the ultimate ideal of being buried within the shade of the holy shrine of Kerbela, where Hussein, Ali's son of tragic memory, lies buried. Many go to that spot to die; others are carried thither by pious relations after death. Those who perform the posthumous journey, however, have a burdensome condition laid upon them by the Turkish sanitary authorities — they must have been dead three years! So it comes about that a temporary resting-place has to be found for them; and here, by Karûn's bank, beneath the secondary shadow of the poor old saint, the little colony of pilgrims in purgatory wait patiently beneath their humble huts of reeds.

The country I am now describing was the scene of Layard's 'Early Adventures.' "Adventures" seems almost too mild a term for the amazing life he led among the cut-throat tribes of the Bakhtiari, Lurs, and Arabs; his wanderings in disguise were as daring as any of Sir Richard Burton's Arabian travels. It was between 1840 and 1842 that the future discoverer of Nineveh came to Arabistan. He was twenty-two years old, and, tiring of work in a London solicitor's office, had decided to try his fortune at the Bar in Ceylon. He conceived the astounding project of making his way thither overland. Passing through Constantinople, Syria, and Bagdad, Layard reached the Bakhtiari country at the moment when one of their principal chiefs was attacked by the Shah's army under a most bloodthirsty and unscrupulous eunuch known as the Matamet. The chief was forced to leave his mountain fastness and flee for his life into the plains of Arabistan, to find refuge with his friend the Sheikh of Mohammerah. The Matamet and his army pursued; and Layard describes how the terrified Arabs broke down all the dykes and irrigation dams, so as to flood the country against the invaders; pulled to bits their huts of reeds (such as they still live in to-day), and made of the *débris* rafts on which they embarked with their families and what few goods they could take. Layard himself, alone and helpless—he had been stripped by brigands shortly before—managed to build a small raft for himself, and, joining the endless flotilla drifting down the river, at length reached comparative safety in the Sheikh's camp.

In those days the tamarisk, which grows in a thick tangle in many places along the Karûn's bank, was a favourite haunt of lions. The following is Layard's account of a quaint popular belief about the King of Beasts: "The lions are divided into Mussulmans and Kafirs or infidels. The first are tawny, the second dark-yellow with a black mane. If a man is attacked by a Mussulman lion he must take off his cap and very humbly supplicate the animal in the name of Ali to have pity on him. The proper formula is, 'O cat of Ali, I am a servant of Ali. Pass by my house (*i.e.*, spare me) by the hand of Ali!' The lion will then generously spare the supplicant and depart. Such consideration must not, however, be expected from a Kafir lion."

The natives are nowadays spared the necessity of such a momentous discrimination, as it is now ten years or more since the last lion was seen in this part of the world. Their quondam prey, the wild pig, live, in consequence, a safe and uneventful life on the river's bank. These, with the shy herds of gazelle which occasionally appear on the horizon, and the ubiquitous jackals whose howling makes night hideous, are almost the only four-footed beasts to inhabit this arid land. Birds, on the contrary, are very numerous—snipe, duck, partridges, and long-legged cranes, and, above all, sand-grouse.

The time of our journey through the desert was the sand-grouse flocking season, and one could ride for hours watching their amazing manœuvres in the sky. On the horizon would appear what looked for all the world like the thick cloud of smoke streaming from an express train. Suddenly the cloud condensed into a solid mass, and an instant later a point shot out of the mass into the sky like an exploding rocket, leaving a wedge-shaped train behind. The next minute a change in the direction of the flock's flight would make it vanish as if by magic, only to reappear farther along the horizon and commence its strange evolutions over again. How many scores of thousands of birds go to make up one such flock, and how they all find food, are questions which must puzzle the most learned ornithologist. They say the beat of wings is quite deafening when a flock of sand-grouse is still two miles away; but for this I could not vouch.

On the third day out from Mohammerah, when the emptiness of the landscape was beginning to grow oppressive, an indistinct whiteness which had been long visible on the northern horizon gradually resolved itself into the great snow-clad Bakhtiari range. Its 12,000-feet peaks formed a dazzling barrier between the plains across which our caravan was slowly crawling and the great tableland beyond, which is the real Persia. The hill tribes which live among these mountains, Lurs and Bakhtiaris, are among the wildest tribes in Persia, and the country has seldom been penetrated by Europeans, its chief explorer of recent times being Captain Wilson himself, the British Deputy Commissioner.

That night we camped at Umm el Tummair, "the Mother of Date Syrup," one of the rare villages dotted on or near the Karûn's banks. The method of nomenclature, of which this is a sample, is, of course, a favourite one among Arabs, who apply it indifferently to persons, places, and animals. The names they give are often delightfully expressive, such as "the Father of Long Noses," signifying a snipe. Applied to human beings they are apt to verge on the personal, as the following instance may show. The Sheikh of this particular village wished to inform us that he had seen one of our foreign colleagues who had gone on ahead of us by the same route. Our colleague was a little conspicuous by his corpulence, and the Sheikh (meaning no offence) expressed himself thus: "This morning the Father of Bellies brought honour to our village."

From this point we finally parted company with the Karûn and struck across to the Kerkha, which here is only a short march away. The Kerkha is the modern name of the ancient Choaspes, whose water—though having nothing particularly tasteful to a European palate—had such a vogue among the old kings of Persia that, even on their remotest campaigns, they refused to drink any other, but had it brought to them daily in golden jars carried across the length of the Empire by relays of horsemen. The river rises near Kermanshah, meanders in a series of inconsequent windings to within a little distance of the Karûn, then turns northwards again, and ends by losing itself among the marshes. During the last part of its course it passes near a place called Howeiza, where some eighty years ago it gave a remarkable display of the fickle nature which it shares in common with all the streams of Mesopotamia. The river at that date passed through the town and watered the fertile lands around. Excellent crops were raised, and Howeiza was a very flourishing town of 30,000 inhabitants. One fateful day, however, the good folk of the place woke to find themselves left high and dry—the river had, in the night, abandoned its old bed and taken to another some miles away. Deprived in this way of the sole cause of its prosperity, Howeiza rapidly declined; in a short time the population had dwindled to a tenth, and nowadays the place is little more than a village.

At the point at which we reached it, near the

village of Kut Said Ali, it was about 100 yards across and very deep. Fortunately we were still in the land of *belems*—though the rudely-built pitch-covered specimens we found here were but poor counterparts of the white, spick-and-span boats at Mohammerah. A small fleet of them ferried us and our baggage across, the horses were towed behind, and the mules, stripped for once of their pack-saddles, were herded together at the top of the bank and driven pell-mell down the steep slope into the water like Gadarene swine, where, finding all retreat cut off, they bravely struck out for the other bank. Some camels followed, tied head to tail in a string, and wearing a look of, if possible, even more abject despondency than they have on dry land.

A family of *seyyids* of great repute live in this neighbourhood. Those *soi-disant* descendants of the Prophet enjoy a position somewhat analogous to that of a country gentleman of fifty years ago; they do no work, and are supported in considerable luxury by the common herd. The head of the family came to call on the British Commission in the course of the afternoon, bringing with him his ten-year-old son, a particularly handsome and jolly-looking boy. Our kit was not yet unpacked, and no cigarettes could be found to offer to the visitor, so the Commissioner's cheroots had to take their place. The *seyyid* took one, eyed it dubiously, and lit it, but evidently found it not at all to his liking. Being too much of a gentleman to commit such a breach of manners as to throw it away, he solemnly handed it on, after a few puffs, to the little fellow squatting by his side. The urchin was still gleefully puffing at his cigar five minutes later when his father took leave,—we missed the inevitable sequel!

A range of low hills lay behind our camp, from the top of which one had a magnificent view of the country lying away to the north. The Kerkha winds through it in a hundred graceful loops and curls, and waters miles upon miles of splendid rolling country. Yet as far as the eye could reach the only sign of human handiwork was a solitary white tomb. At first it seems incredible that such a tract of land should be left absolutely desert—a sort of No Man's Land; but the reason is not far to seek. All this region is the happy hunting-ground of the Sagwand and the Beni Lam, two of the most predatory tribes in Asia. The Beni Lam have their headquarters down by the Tigris, near Amara, and were the chief Arab tribe to throw in their lot with the Turk against us in the present campaign. Seventy years ago they enjoyed the following unsavoury reputation among their neighbours, as quoted by a European traveller through their country: "The Beni Lam are not Arabs but Kafirs, who neither respect the laws of hospitality nor behave in any way as good Mussulmans. They are as treacherous as they are savage and cruel, and would cut the throat of a guest for a trifle." The cap still fits.

These amiable tribes live almost exclusively by plunder and rapine, robbing and harrying the poor villagers, till the latter live in a state of constant terror of them. The consequence is that a huge tract of extremely fertile country, which a little labour in irrigating would make as productive as any in the world, remains barren and untouched. Only a week before our arrival the Beni Lam had swooped down on the flocks of Kut Said Ali and driven them off. They had avoided actually killing the boys who were guarding the animals, because they do not like the inconvenience of a blood-feud, but in order to delay their getting home to give the alarm, they had taken the poor little chaps and flogged them till they could hardly walk.

Once or twice we ourselves were taken for a raiding party by shepherds grazing their sheep in the plain. As soon as they caught sight of us, they would start to drive their flocks helter-skelter towards the nearest village, shouting with all their might and firing off their guns to attract help from the villagers. On the first occasion I galloped after them to reassure them—overtook them, and found myself looking down the muzzle of a loaded ·450—after that I left the reassuring to some one else to do!

One day's march from the crossing of the Kerkha brought us to the edge of the Khor-el-Azem. The caravan now turned off to skirt round the edge of the marsh, but I was lucky enough to be able to send my horse round with the main party and travel myself for two days by *belem*. The marsh scenery is wholly unlike anything I have seen elsewhere, and hardly less unique is its population of queer amphibious beings who live among their swamps, isolated from the outside world, and earning a meagre livelihood by growing rice and fishing.

I find several notes in my diary, made during our lazy progress down the stream, which I think will best give the impression of what the journey was like. The first is dated—

Feb. 22, *noon.*—We are floating down the Kerkha in a *belem*. Rather a tight fit—nine persons inside! It is a primitive tubby variety of the *belem*, with a tremendously long tapering bow curving back so as to give to the boat almost the outline of a Viking's ship. It is divided by two thwarts, and the first-class accommodation amidships is occupied by the Colonel, myself, and an Arab *seyyid* with a bright red henna-stained beard and a limitless capacity for making unpleasant noises and soliciting presents. It's drowsy work, drifting down on the stream with occasional bursts of frenzied energy when our *belemjis* churn the water with their paddles for a few minutes, and then relapse into idleness and droning Arab songs. The boats we meet coming up-stream loaded with cut reeds are tracked by a man on the bank, while his pal sits in the stern and steers the bow of the boat out from the bank. These marshmen wear very few clothes and are burnt nearly black.

2 P.M.—After lunch and a shoot on the bank, which is full of frankolin and hares, we are again paddling down-stream. The boatmen here seem never to have discovered the possibility of paddling one on each side of the boat; they give a few strokes together on the bow side, then swing over and do the same on stroke side; result—a very zigzag course and frequent sprinklings for the passengers. The river banks are populated by innumerable tortoises, who sit and crane their necks at us as we go by. There are solemn cranes standing sentinel here and there, and kingfishers, some of them blue, some black and white, flitting over the water. We have just passed the queerest group of birds sitting on the bank. There were about half a dozen of them, very grave hunchback creatures, rather like small penguins, but without any of the penguin's cheerfulness. They reminded me of a party of Scotch elders at a funeral, as they stood there silent and motionless and wearing the most dejected air imaginable. One had a gorgeous greeny-blue back, but the rest were in sober grey —perhaps his *harim.* The Arabs say they are called *wag*, and only the Bedouins eat them.

Later (*in Camp*).—We have arrived at Bisaitin, one of the biggest of the marsh villages. It stretches as a single row of huts for miles along each bank of the river, with side streets at intervals on canals leading off the main stream. The huts are long and narrow, the walls consist of bundles of reeds about six feet high, partly sunk into the ground and covered with a "barrel" roof of reed mats; they look very unlike any house you ever saw, being just a thick sausage with a big tuft sticking up at each end where the uncut tops of the reed bundles are allowed to hang over above the roof. Each village has one or two mud palaces where the big-wigs live, and all the life of the place goes on on the waterways. Half-way through the village we branched off down a side canal, came to a place where it broadened out into a shallow lake, and found the camp already pitched at the water's edge, so sailed right up to our own front door.

23*rd*, 9 A.M.—Back in our *belem*, this time with W. We have managed at last to dislodge Red Beard into the bows and can stretch our legs. For an hour we slipped past an endless succession of reed-huts, and crowds of staring Arabs and naked children lining the bank, very interested in their first view of a European—for this bit has never been travelled along before, as far as we know. Now and then we overtook another *belem*, and had animated if somewhat unintelligible conversations with the occupants. They start by saying, "There are some terrible fellows up the river where you are going; they'll cut your throats and bash your heads in" (actions to suit). We reply by pointing to our guns, and they clap hands and applaud our bravery. Then a large hubble-bubble is passed on board for our *belemjis* to have a couple of sucks at while we inspect their fish tridents.

A mile back the Kerkha abruptly came to an end amidst impenetrable reeds, and seeing no way out, we thought our guides had deceived us, but the marshmen turned out in force and pulled and shoved us over a bar into a hidden canal about five feet wide and full of other *belems*. It was a miniature Boulter's Lock, and we shoved and cursed and laughed until at last we got through, and are now meandering along a vague channel among the reeds which looks as if it would come to an end at every corner. We have taken on board a fine fish which some men we passed had just speared. I forgot to mention that to-morrow's dinner, in the shape of a live sheep, is tied to my thwart, and occasionally butts me in the back or nibbles my shoulder.

10 A.M.—We are still poling and paddling along this extraordinary channel, only a few feet wide, with a sharp turn every few yards, and an impenetrable wall of rushes six feet high completely shutting us in, so that all one can see is the sky and a few yards of water ahead and behind. Two other *belems* follow us, paddled by men cross-legged in the bows and crooning Arab love-songs. One hears birds but sees none, and it's roasting

hot. I doubt if a white man could live here an hour at midday in summer.

W.'s wild footman is sitting behind him, a man from the mountains, who illustrates the delightfully characteristic Persian trait of telling any sort of lie to please his listeners. Whenever Red Beard says, "It is so many miles to so and so," or, "There will be good shooting at to-night's camp," the Shatir encores, "Yes, just so many miles," or, "Splendid shooting." Of course he knows we know he has just come from Luristan, and has never been within two hundred miles of the place before, but that doesn't worry him. I have just overheard W., in reply to an inquiry, explain in his best Persian that gamooses *do not* walk about the streets of London!

Camp at Umm Chir, 3 P.M.—We paddled for another mile through the reeds, then came out into a lagoon thick with water-fowl of half a dozen different sorts, with the desert sloping down to it beyond. This is the limit of the marsh, so we waded ashore through the shallows and came on a mile into camp.

CHAPTER V.

THROUGH THE LAND OF ELAM (2).

WHAT I may describe as the second stage of the frontier began from Umm Chir. We now had before us 200 miles of pure desert, so utterly bare that we were destined to travel 150 of them without seeing a trace of any human habitation. The way lay along the edge of the great Mesopotamian plain, a bare arid track, watered only by the scanty streams which come off the Eastern watershed of the Luristan mountains and flow down to the marshes which fringe the bank of the Tigris. A fair description of this region was left by a member of the Frontier Commission of 1849-52 in the following words: "It is extremely difficult country to travel through on account of the absence of habitations, the danger of attacks from the Beni Lam Arabs from the Tigris side, and the Lurs from the directions of the mountains, and the brackish and pitchy water which is hardly fit for use in winter and is quite undrinkable in summer. This country has no boundaries but the Tigris on the west and the Luristan mountains on the east.

The banks of the river are inhabited by the Beni Lam, the interior of the mountains by the Lurs; the strip of desert serves as a line where these peoples meet, a kind of neutral ground on which they sometimes fight each other and sometimes fraternise, but they are always ready to plunder a traveller, so that caravans never dare to travel these regions." The old Commission seem to have had a wholesome respect for these brigand tribes. They—that is the English, Russians, and Persians—took provisions for sixteen days, loaded 107 camels with forage, armed all their servants and followers, "so that the whole thing had the appearance of a military expedition," and practically made a dash for it. Troops of robbers, they relate, stealthily followed the caravan, and there were alarms and shots in the camp almost every night. So harassed were they that they were only able to make a very rough map of this part, calculating the distances by the length of a horse's step. The result of all this was that from near Umm Chir to Mendeli not even an approximate line of frontier could be laid down beforehand by the Ambassadors and the Grand Vizier at Constantinople, and so there fell to our Commission the work not only of its demarcation but also of the delimitation itself. The only principle to guide the Commissioners in settling a line was the *status quo* frontier. But this did not help very materially in a region where there was scarcely a living soul to tell where the recognised line was, even if, as was more than doubtful, a line ever had been recognised. Some shepherds were at last found feeding a meagre flock of goats in the desert, and were hailed with alacrity before the Commission and questioned.

"Whose country are you feeding your goats in?" they were asked.

"Allah's," they replied, nor could any other information be got out of them.

The Turks and Persians, advancing very divergent claims, could come to no understanding about the matter, so eventually Anglo-Russian arbitration was invoked and the frontier line fixed for the next 80 miles ahead.

All this, of course, took some days, during which we were able to explore the sporting possibilities of the neighbourhood. Game were fairly plentiful and varied. Large herds of gazelle were reported by the Indian Surveyors returning from plane-tabling, but our pack-mules, scattering for miles around in search of grazing, and accompanied as always by their bells, spoiled our chances of venison for dinner. Then there were hares, frankolin, sand-grouse, and bustard, the last splendid birds, as big as young turkeys, and most succulent eating, but almost as wild and unapproachable as the gazelle. Great numbers of wild pig lived near the edge of the marsh, spending the daytime in the marsh itself and returning in the evening to the dry broken country behind our camp. My first introduction to *Sus persica* was of a quite dramatic nature. We had just finished lunch, and I was standing talking with a Russian officer in the camp when a clatter of hoofs made us look round. There was a mounted Arab galloping wildly through the camp and driving two large boars in front of him. The pair came trotting in among the tents with such a nonchalant air that I had for a moment the absurd idea that they must be tame animals. Not for long, however, for as soon as they reached the Arab encampment all its occupants turned out with yells and guns and commenced a tremendous fusillade at point-blank range. The luckless beasts had to run the gauntlet for about 50 yards under a withering cross-fire, but came out unscathed at the end, and got safely away into their native marsh. Nor, *mirabile dictu*, were there any casualties among their reckless aggressors.

One day the shooting-bag included a wild cat, a splendid specimen, measuring just over 4 feet from tip to tip, whose skin went, with our other specimens of Persian *fauna*, to the Bombay Natural History Museum. Another martyr to science was an enormous horned owl, who measured 4 feet 6 inches across the wings. Lastly, I must not forget to include those most ubiquitous of subterraneous pests, the jerboa rats. Individually they are the most fascinating of little creatures, with their big eyes and preposterous long tufted tails, but collectively they are nothing short of a public nuisance. The whole country was honeycombed with their burrowings, made worse by the smallness of the entrance holes, so that you could never tell at what moment the ground would give way under your horse's hoofs; luckily it was soft falling.

The frontier, as finally settled, was to follow for thirty miles or so the course of a dried-up river-bed, called the Shatt el Amma, or "blind stream." This seemed straightforward enough, but the sequel exemplified well the troubles of frontier making. The marsh by whose edge we were encamped had originally spread a good deal farther west, and in receding had left behind some square miles of country which looked like nothing on earth so much as a dislocated jig-saw puzzle. The bits of the puzzle were represented by tussocks of all shapes and sizes, about 18 inches high, and formed of the roots of long-dead reed clumps; in and between them ran a perfect network of narrow, deep-cut channels, and somewhere through the middle of all this was the Shatt el Amma. However, the survey officers managed somehow to unravel the problem, though several times the river-bed was so completely lost that the only thing to do was to put up a frontier-pillar where it disappeared and then cast ahead, to pick it up again perhaps a mile farther on.

Our next halting-place was to be the river Douerij, but before we reached it we were met by the news that the Beni Lam had succeeded in stealing three troop-horses belonging to a small party of *sowars* who had gone on ahead. This was our first, but by no means our last, experience of the attention of this clan of accomplished robbers. Before we were quit of their country the toll they had levied on the joint four Commissions included a dozen or so mules, a horse (the troop-horses were eventually recovered), four camels, loads and all (luckily for us, and much to the disgust, I should imagine, of their captors, these consisted chiefly of cement for boundary-pillars), and, last but not least, the Russian Commissioner's uniform frock-coat, stolen brazenly from his own tent.

The Douerij, when we reached it, turned out to be a fast-running stream fifty or sixty yards wide flowing at a level far below the desert between precipitous mud cliffs. We camped by a ford, and crossed in pouring rain next day. The ford was nearly five feet deep, and the *charvadars* were only induced to attempt the crossing by the persuasive influence of our deputy transport officer, who, clothed in a streaming shirt and much fine language, and armed with a convincing shillelagh, refused to argue the matter. A serviceable raft was constructed of blown-out water-skins and sowars' lances which took the tents and heavier loads, while the men portered over the rest. The river meanwhile was rapidly rising and the rain continued to fall in torrents, so, forsaking the river's edge, we camped on a broad ledge half-way up the further bank. All that day and the following night the storm kept on, and though by the morning the sun had re-asserted itself, the river was now a swirling, raging flood of chocolate-coloured water covered with foam, uprooted bushes, and the *débris* of trees brought down from the mountains. We were quietly taking photographs at the river's edge when a shout from the camp brought us back, to find that the water had encroached unobserved from the rear and was on the point of flooding us out. The bank behind our ledge was a good ten feet high and nearly perpendicular, so there was no time to lose. The tent's ropes were thrown off and down came the big "Hudson Raotis," each of them a full five mules' load on march, with a run into several inches of mud and water. There was no time to even detach the flies, but thirty men pulling for all they were worth from the top of the bank just hauled them up soon enough to avert disaster. Half an hour later our quondam camping-ground was only distinguishable from the rest of the river by a few tops of bushes bending to the flood. Our Persian friends, camped on the other bank a little way back from the edge, had had a good laugh at us, but it was soon to be our turn to smile. As the volume of water increased, the soft mud cliff, pounded against from below, began to undermine, and presently huge masses of it were crashing periodically into the river with a noise like thunder. The Persians packed up hurriedly and took to the hills. By tea-time, however, the tables had turned on us again. Our new camp was on the general level of the plain, and at first it seemed preposterous to suppose we were still in danger, yet the river rose inch by inch and was now over twenty feet above its level of two days before, and had changed from a fair-sized stream to a river almost the size of the Thames and three times as rapid. By five o'clock there was only three feet of "freeboard" left, and the rate of rise was six inches every half-hour, so we resigned ourselves to the inevitable, called in the mules, and trekked to the nearest hills. We got little sleep that night for the booming which announced every few minutes the collapse of another few hundred tons of the river's banks. On the third morning the water-level, which had reached during the night to within two inches of the desert, began to subside, and we ceased to feel like Noah's wicked contemporaries driven higher and higher each day by the flood. The river was still tremendous—the most irresistible, devastating thing, I think, I ever saw; but as no human beings live near its banks it did nobody but ourselves any harm, and being on the right side of it, we could afford to speculate with interest how our friends on the yonder bank would set about to negotiate the crossing. For two days we were entirely cut off from intercourse; but on the third, when the waters had abated to reasonable proportions, there arrived a large *belem* borne on the backs of two much-enduring mules which the Turks had sent down to Amara for the purpose, and on the fourth day we were once more united.

The result of the flood might have provided an interesting study for a geologist. So tremendous had been the undercutting action of the current, that in places the river had demolished a strip of the desert twice or three times its own breadth, and so created an entirely new channel for itself. The sight made one speculate on the accumulated effect of such floods as this over the space of a few æons. The analogy of Noah, by the way, was a closer one than might at first appear. There is a theory, propounded, if I am not mistaken, by Sir William Willcocks, that the flood of the Book of Genesis actually consisted in the water rising (as it so nearly did in our case) above the general desert level—a condition of affairs which the great engineer believed would be the natural result of the simultaneous occurrence of the following phenomena: a big spate on the Euphrates, the same on the Tigris, and a strong south wind blowing for several days on end up the Persian Gulf and so banking up the waters of the combined rivers. There is even a slightly raised eminence near Bagdad which, it is claimed, would be the only part of Mesopotamia left unsubmerged, and so may reasonably be identified with the Mount Ararat on which the ark stranded. It is perhaps unnecessary to add that the 17,000-foot mountain which bears that name on the map was so christened in comparatively recent times by devout but imaginative Armenians.

We were kept in continual remembrance of our adventure on the Douerij for some days after we had left it behind us for good. Somehow during the bustle and confusion of our sudden and unpremeditated flittings the flour and the kerosene had found themselves in close proximity to each other—and had joined forces. No matter how sturdy an appetite one brought to one's dinner, the first bite of *chupattie* banished it for the rest of the meal. Let me utter to all travellers a most solemn warning, prompted by the bitterest experience, that however excellent the two constituents may be, each in its own sphere, as a combination they are *not* a success.

The Luristan mountains, as we gradually converged towards them, grew more and more forbidding, their outline against the Eastern sky becoming a grim chain of rugged peaks and precipices. At the Douerij we were still many miles away from the main range (of which this section bears the name of Kebir Kuh, or "great mountains"), but had reached the low sandy foothills which encroach far into the plain, and formed, from now on, the frontier line.

We were already in the month of March, and the hollows among the hills were gay with the spring flowers which crop up in places even in the most uncompromising deserts at this season of the year; gentians and cuckoo-spit, and a flower like a large scentless cowslip lined the bed of dry *nullahs*, and every here and there a thick carpet of anemones added a glorious splash of crimson.

It was about this time we fell in with old Ibrahim. He appeared in camp one day apparently from nowhere—for we must have been a good hundred miles from the nearest village—and never left us again till the journey's end. His own account of himself was as follows. He was an Indian born at Bombay, and had accompanied his father to Bagdad at the tender age of five; there he grew up and prospered till, like many of his fellow-men in colder climes, he came to ruin over horses. A speculative shipment of ponies to India went wrong and left him penniless, so he became a pedlar, and had spent a lifetime hawking his wares from Bagdad to Erzeroum. Having passed many times, as he said, along the route which lay before the Commission, and knowing the language and customs of the different tribes we should meet, he proposed his services as a sort of walking Baedeker, and as he was a hardy old fellow despite his eighty years, and asked for nothing more than his food and footgear, he was duly enrolled. Later on he so far justified his claim to intimacy with the tribes, that he turned out to have married, at one stage or another of his peddling career, a wife from nearly every one of them in turn!

His *début* as a guide, however, very nearly

sealed his fate as far as the Commission was concerned. We had camped about twenty-five miles short of the river Tyb, our next place of rendezvous, and two of us, Captain Wilson and myself, set out with a small party in advance to locate the ford. Ibrahim, asserting that he knew exactly where it was, came as guide. A few miles out from camp we caught sight, to the south of us, of the ruins of Shahriz, one of the multitude of dead cities scattered throughout this corner of the world, the birth- and burial-place of so many ancient civilisations. Nothing is to be learnt of their history, save in the case of the few great excavated sites like Susa, Babylon, and Nineveh, and the traveller, unable to so much as guess to which civilisation each belonged, can only wonder at their size and the utterness of their destruction. When, if ever, these regions emerge from beneath the pall of their present desolation, what a playground for archæologists they will be! There is, I think, a special fascination in walking the silent streets of a dead city, and though we knew from Layard that there was nothing above ground to see, we left the caravan to continue its way, and branched off to visit the site. We found there the well-defined remains of a town about a mile in diameter, surrounded by a rampart and a moat. Among a sea of shapeless heaps the lines of the principal streets were still traceable, leading up, in the centre of the town, to a mound of extra large size which covers, no doubt, the ruins of some sort of citadel. A profusion of scraps of broken pottery covered with a rich blue glaze, lying scattered everywhere, were all that time and the wandering Bedouin had left to bear witness to man's handiwork.

We rode away from the city of the dead, intending to intercept our caravan a few miles ahead. It was only after an hour's galloping, however, that the familiar line of creeping black dots appeared, emerging from a shallow depression in the desert a long way off. Presently they were lost to view again behind some low bushes, whence, to our surprise, they did not reappear. We had come to within a few hundred yards of the spot, when a shot suddenly rang out from the bushes and brought us to an abrupt halt. A head looked out a moment later, and its owner having satisfied himself apparently of our pacific intent, he and his companions emerged from their ambush. What we had taken for our caravan proved to be a party of four or five Lurs returning to their native mountains from a "shopping expedition" in the Tigris valley—of a very up-to-date nature, to judge by the huge gramophone trumpet which was balanced precariously on the top of one of their mules' loads. The shot they had fired was merely the usual greeting accorded to a stranger in their part of the world, and implied no particular ill-will on their part; in fact, once it was established that we were not robbers ourselves, and were too well armed to be conveniently robbed, our relations became of the friendliest.

Taking leave of our Lur friends after a little mutual gossip, with greater cordiality than had signalised our meeting, we hurried on to the Tyb, where we found the ford but no caravan. The latter had been led hopelessly astray by the trusted Ibrahim, and did not make its reappearance till two days later, leaving us foodless, and with no alternative than to face the weary journey back to camp again. It was on the return ride, just as it was growing dusk, that we came upon a large herd of a hundred or more gazelle, who gave a touching proof of their ignorance of the human genus by quietly trotting along in front of us for a mile or so, within the easiest range—an ignorance which, I am glad to say, we were too weary to dispel.

The Tyb, at the point where we intersected it, is a very different affair to the sullen Douerij flowing through its deep-cut trench in the desert. Fed by the snowfields of the Kebir Kuh forty miles away, it arrives at the low line of hills which form the last outposts of the range in all the vigour of youth, and flows impetuously through the channel it has hewn for itself before spreading out lazily in the Mesopotamian plain.

The hills here are composed of a curious reddish mud — hence their name of Jebel Hamrine, or the "Red Mountains"—which lends itself particularly well to the plastic forces of the stream. You find yourself in a miniature reproduction of Swiss mountains — little Jungfraus and Matterhorns 100 feet high surround you, so steep that they would be quite unscaleable if it were not for the peculiar consistency of the red mud which gives your feet an excellent grip. The river runs in a deep gorge, whose sides have been sculptured in places into the semblance of gigantic architectural masses. One such mass, in particular, situated at a bend of the river, has been carved by nature into the form of a great cathedral embedded in the face of the cliff, the apse and part of the nave, with roof complete, emerging with almost perfect truth of outline from the sheer wall of clay. These cliffs are the haunt of wild pigeons, and happening in the evening to have climbed to the top of a hill in search of them, I was rewarded with one of the most vivid colour effects I ever saw. The sun, just on the point of setting, broke for a moment from beneath a bank of heavy clouds hanging above the horizon. The moment it did so the mud peaks and cliffs all round were transformed by a quite indescribable glow of red gold, emphasised the more by the blackness of the valleys between, while at the same time the big mountains beyond were flushed by the softest purple light. A minute later the sun dipped below the horizon, and the colours faded as quickly as they had come, leaving nothing but dim grey landscape behind. It gave one almost the sensation as if a coloured limelight had been thrown for an instant across a darkened stage and then as swiftly withdrawn.

Beyond the river the hills open out into a broad bay of terraced grass slopes, and there our tents were pitched, while the noisy mules and followers were banished to humbler planes below. The weather meanwhile was perfect and the shooting very fair. For wayworn travellers like ourselves it seemed an ideal resting-place, yet on the third day the Commission fled from this delectable spot as from a city stricken with the plague. The cause of our precipitous flight was none other than the Tyb itself. Its rather peculiar flavour had been noticeable from the first, but it was not until half the camp was sorely stricken, and the doctor analysed the water, that it was known for what it was—an exceedingly potent solution of Epsom Salts!

The question of water is, of course, a primary one for travellers in the country we were passing through, and one's degree of comfort or discomfort in camp varies in pretty direct ratio with the goodness or badness of the water supply. For an example of possible extremes, I need not go further than our experiences at our next two camping-grounds after we left the banks of the "father of Epsom Salts." The first camp was at Kara Tepeh, a huge solitary mound in the flat desert, quarried with jackal's holes and Bedouin graves, and hiding Allah-knows-what long-forgotten ruins. At the foot of the mound there are half a dozen shallow wells. A thick green scum is all you can see when you look down them, but skilful manipulation of a bucket will bring up a small quantity of thick fluid of the precise colour and consistency of *café au lait*, and only recognisable as water by the presence in it of millions of tadpoles. We did not prolong our stay at Kara Tepeh! The very next camp carried us, as I say, to the other extreme. We were still on the edge of the same arid desert, though a little nearer the hills, and close by the camp, hemmed in between high cliffs, ran the jolliest little brook imaginable, rollicking down over a pebble-strewn bed, and spreading out every here and there into a quiet deep pool full of darting fish, and clear as crystal. Such luck is, however, rare, and even here we should probably have found nothing but a dry *nullah* if we had come a few weeks later.

It may not be amiss, before closing this chapter, to give some account of our caravan and our mode of life in the desert. From the first, though the hot weather had not yet begun, we adopted a summer time-table, getting up on marching days an hour before sunrise. During January at least this meant a shivering breakfast by either moon- or candle-light in the open, for the mess-tent, packed up over night, went on ahead with the *pishbar*, a small advance party which, following Persian custom, preceded the main caravan by an hour or so. While we consumed our coffee and *dhal-bât* the mules were loaded up, a fairly tedious business even when confusion was reduced to a minimum by each mule being labelled in large letters with the initials of the owner of his load. There are as many different ways of loading a mule as of cooking an egg, and the particular system favoured by Persian muleteers is, I believe, peculiar to themselves. Every mule has, as part of his equipment, a length of stout woollen stuff about three feet by nine, with cord-laces along the ends and corresponding eyelet holes. The beast's load is divided up into three parts, two equal ones of roughly 80 lb. each, and a third, preferably some small and compact object, of about 50 lb. The long piece of stuff having been spread out on the ground, the two bulkier lots are laid on it, one on each side of the middle, and the ends turned over and laced up in such a way as to form a sort of double valise. Then comes the tug-of-war. The *charvadar* ties his animal's bead-rope round his own waist, and, aided by a comrade, hoists the load up against the flank of the mule, who is encouraged by loud and violent objurgations to lean his weight against it, thus allowing one-half to be toppled across his back so as to fall on the farther side. This part of the loading process obviously implies the goodwill and co-operation of the mule himself, and it is really surprising to see how seldom the appeal to his better nature fails. When the two loads have been well adjusted in equilibrium, the third part—the *sar-i-bar* as it is called—is hoisted on top between the other two, and a long woollen girth having been thrown over the loads, passed under the beast's belly and hauled as tight as the *charvadar* can pull, the mule is ready for the road. Despite the unadaptable qualities of such objects as tin baths, helmet cases, and 9-foot tent poles, it was comparatively rare for a mule to shift his load.

The caravan on march stretched for three or four miles across the plain, split up into groups of ten or twenty mules, with a sprinkling of *sowars* along the line to ward against attack, and a rearguard to encourage stragglers.

In the cool dawn it was very pleasant to give one's horse his head and gallop from end to end of the whole line, the chiming of the mule-bells in your ears waxing and waning as you overtook group after group of the plodding beasts. It was amusing, too, to watch, as one passed, the queer variety of types which went to make up the long column—the easy-striding *jillou-dars* in their much-patched, long-skirted coats, shawl cummerbunds, and full bell-mouthed trousers; the more important mule-owners on their neat Arab mares, looking particularly light by the side of the Lancers' big walers and Indian "country-breds"; the *kholassis* in business-like khaki and puttees, carrying fragile theodolites, tripods, and other survey instruments; the green-liveried mess-servants, the Goanese butler and his colleague the cook, the local Arab guide, the *shâtir*,—but the *shâtir* deserves a paragraph to himself.

Persian custom prescribes that a gentleman of rank on his travels shall be preceded on all occasions by a *shâtir* or running footman, and the British Commissioner was accordingly provided with this functionary. He was a Lur of fine physique but singularly mournful countenance, and, arrayed in a frock-coat braided with scarlet and gold, he marched through the desert ahead of his master's horse with all the dignified solemnity of an Esquire Bedell conducting the Vice-Chancellor to Great St Mary's. Anywhere else his fantastic figure would have struck a note of comedy; but in Persia one loses one's sense of the incongruous, so universal is incongruity.

But I must complete my interrupted list of fellow-travellers by mentioning three last characters: "Mick," the doctor's puppy, who, still too young for long marches, rode upon his own mule perched up in front of his master's bearer; "Azaphela," the most diminutive of dachshunds, who travelled in a little cage carefully designed by her owner, Captain Dyer; and last of all, two small white cocks (adopted in infancy by our O.C. Escort), whose lusty voices and martial ardour at the end of a long day's march were not a whit impaired by their having performed the journey in a bucket slung over a mule's back.

The average length of our daily march was twenty miles, and as this meant six hours going without a halt, it was refreshing to find the mess-tent and tiffin basket, carried on ahead by the *pishbar*, waiting to welcome us at the journey's end. Our base of supplies, I should mention, so long as we were in Mesopotamia, was the Tigris. As we worked northwards, provisions and forage were collected successively at Amara, Ali-el-gharbi, Kut-el-Amara, and finally Bagdad, and our transport officer spent his time in travelling backwards and forwards with supplies between these places and our line of march. The arrival of Captain Dyer in camp, accompanied by the week's mail and a dozen or so of bottled beer, were the red-letter days in the Commission's existence.

CHAPTER VI.

THE WÂLI OF PUSHT-I-KUH.

IN 1810 two officers of the British Army, Captain Grant and Lieutenant Fotheringham, adventured on a Political Mission into the then unexplored district of Pusht-i-Kuh. They were received by the Chief of that country, and entertained to dinner, but in the course of the meal were fallen upon from behind and bound. The Chief then had them led to the top of a high cliff overlooking the valley of the river Chengouleh, and there offered them the choice between conversion to Islam and death. Both the officers unhesitatingly chose the second alternative, and were hurled over the edge of the cliff, to be dashed to death on the boulders below. Since that event only three or four Europeans—one of whom, at least, is supposed to have met with the same fate as the two Englishmen—have penetrated the country. It is a wild strip of mountains, about 160 miles in length, contained between the highest ridges of the Zagros range and the Mesopotamian plain, and stretching N.W. to S.E. from near Mendeli to the borders of Arabistan. Its name, which may be translated "Back o' the mountains," describes its position from the point of view of Persia beyond the main range. This remote district has been, as far as is known, a quasi-independent State from time immemorial, and thereby illustrates what appears to have been a point of policy with all the kingdoms of this part of the world. The kings of Parthia, Persia, and Assyria alike found it a wise plan—though often, no doubt, making of necessity a virtue—to maintain along their frontiers small but practically independent chieftains, whose loyalty could be more or less assured by gifts of money and honorific titles. They inaugurated, in fact, the modern political principle of "buffer States." Of such was Pusht-i-Kuh, and still, to some extent, it is to-day.

Its present ruler, Ghulam Riza Khan, is the 14th of his line, and governs for all practical purposes as an autocrat. The authority of the Teheran Government is in his case even more shadowy than in that of his neighbour the Sheikh of Mohammerah, though he is recognised as a Persian functionary, inasmuch as he receives an emolument as Warden of the Marches. It is a peculiar custom of his country that quite a large section of the inhabitants (Lurs by race) are in perpetual attendance on their Chief, and accompany him to his various winter and summer residences; being all of them armed, they form a sort of small standing army. Here also, even more so than at Mohammerah, the chief posts of trust are held by negro *ghulams*. These black men, who themselves or whose parents were originally brought into Persia as slaves, attain, presumably by virtue of a degree of faithfulness uncommon among the Lurs, to high positions of authority as bailiffs of the Chief, and intermarry freely with his light-skinned subjects.

The Wâli of Pusht-i-Kuh (for such is his sonorous title, reminiscent somewhat of the immortal heroes of Gilbert and Sullivan opera) was an interested party to the frontier question. In the impregnable mountains which form almost the whole of his domain, there is a gap at the point which we had now reached where the Chengouleh of evil fame issues through the Jebel Hamrine and waters a considerable tract of land, spreading downwards in the shape of a loop towards the Tigris. The Wâli claimed that this tract was cultivated by his men, and so belonged by prescriptive right to Persia; the Turks denied the claim. The rival Commissioners showed little signs of compromise; indeed their mutual attitude was rather naïvely illustrated by a remark made by one of them to a neutral member of the Commission: "Mais ces terrains là nous ne les contestons pas, c'est l'autre parti qui les conteste!" The question was not made the less complicated by the fact that "ces terrains là" had eighty years before been the site of a flourishing small town standing among groves of date-palms. On the death of the Wâli of that day the succession was disputed by his three sons, one of whom, descending on Baksai, the village in question, vented his wrath on the brother who owned it by utterly razing it to the ground and felling all the palms, to the number, history relates, of 11,000.

At the time of our visiting the place not a sign of its former prosperity was left beyond a few ruins of mud-built aqueducts and water-mills, some rotting palm-stumps, and a deserted tomb-mosque gaping to the four winds. Soon after we had pitched camp at Baksai, a present arrived from the Wâli. It took the eminently practical, if rather unusual, form of a large lump of snow; he had brought a quantity down from his mountains, and as the plains were now beginning to stoke up and the thermometer stood high in the nineties, it was quite the most acceptable gift he could have sent. Next day he came himself in full state. His train as it wound across the desert to our camp was, it must be confessed, a trifle suggestive of a circus procession. First came a guard of riflemen, two by two, on foot, then a led horse, the usual sign of rank, followed by the court band, in tattered red uniforms, playing on cornets and a big drum (they were once, it is said, in the service of a Pasha of Bagdad and formed part of the Wâli's spoils of war after a successful engagement with the imperial troops). Next, preceded by his *shâtir*, a resplendent individual in scarlet and gold, came the Wâli himself, a tall bent figure with coloured spectacles, riding a fine Arab horse with a leopard skin thrown over the saddle ; with him were his two sons, and behind rode the son of Salar-ed-Douleh, that arch-rebel and Pretender to the Persian throne, a handsome little boy in a very smart suit, on a horse with gold bridle and trappings. The rear was brought up by a long string of armed horsemen.

We returned the Wâli's visit next day. Arriving at his tent, we were met by two magnificent footmen dressed in full-skirted red coats, braided across the chest like a hussar, embroidered white stockings, and the most imposing hats you ever saw, of the same inverted saucepan shape as the *charvadar's* already described, but of far greater proportions, being about eighteen inches high and twelve across the top, and encircled at the base by a coloured turban. These splendid individuals, each carrying a *bâton* like a drum-major, conducted us to the door of the tent, whence we were ushered by the Master of Ceremonies into the presence of the Wâli himself. The Wâli was installed in a chair at the end facing the door, and placed lengthways down each side of the tent were two ordinary iron bedsteads. These were, it appeared, intended for our accommodation, so we arranged ourselves along them in two rows facing each other.

A funny little group squatted behind our host,

composed of his three youngest sons, the smallest of whom was a little fellow of four, with henna-stained curls and a long green frock-coat, who attempted, with some success, to play practical jokes with the rickety bedsteads during the audience, and the ancient Vizier, a hairy Rip Van Winkle, who, throughout the conversation, croaked hoarse promptings into his master's ear. The only other persons present were the Master of Ceremonies and two grown-up sons, who stood demurely in the presence of their father with arms folded and hands hidden within their sleeves, as Persian etiquette demands.

The Wâli himself had donned a curiously mixed costume for the occasion. He wore black alpaca trousers and patent-leather shoes, with a sort of military frock-coat, the epaulettes of which were adorned with brilliants set in the device of the Turkish *tughra*, or imperial cipher—perhaps of the same origin as his band. He was not very talkative, and conversation was fitful, as becomes an official visit in Persia. He was interested, however, in aeronautics, and asked questions on the subject, whereupon a discourse ensued rather on the lines of that chronicled in the first chapter of 'Eothen,' when Kinglake and the Pasha of Belgrade exchange their views on steam-engines. But the subjects nearest to the Wâli's heart related to his royal brethren (as he doubtless regarded them) on the thrones of Europe. "Who," asked he, "is the Padishah of Inghilterra?" "Jarge," replied our leading Persian scholar, who was carrying on the conversation (this, by the way, was not gross *lèse-majesté*, but merely in strict accordance with Persian pronunciation). The Wâli turned to his Vizier. "Write down Jarge," he said. "And he of Russya?" he turned again to his guest. "Nee-ko-las" was the answer. "Write down Nee-ko-las," to the Vizier. "And what is the name of the Padishah of Allemân?" Our spokesman assumed the air of one racking his memory for some obscure and half-forgotten fact, then replied in dubious tones that he thought it was something like Weel-Yâm. The German Emperor's name does not, I have reason to believe, figure in the Royal Gazetteer of Pusht-i-Kuh.

The arrival of coffee presently intimated that the guests were at liberty to go, a hint trenchantly emphasised by the sharp iron edges of the bedsteads. So with the prescribed bows and salaams to our host and his suite we took our leave.

CHAPTER VII.

TOWARDS BAGDAD.

The Arbitrating Commissioners having ultimately decided on an equitable solution of the vexed question of the Baksai lands, the Commission proceeded on its way. To the north of Baksai there occurs a curious break of level in the desert. An almost perfectly straight cliff, 200 to 300 feet high and 30 miles long, runs out obliquely from the mountains and causes this phenomenon. Some powerful force of erosion is, or has been, at work, resulting in the disintegration of the sandstone of which the edge of the desert is composed. Along the foot of this cliff, for a depth of two or three miles, there stretches a strip of extraordinary formation, a jumble of little hills and hollows so broken-up and confused that it looks from a distance like a bit of very choppy sea which has become petrified.

As there are but three points at which the cliff is scaleable, and we elected to cross at the farthest of these, we pitched our first camp on the hither side. Next morning, starting before daybreak, the caravan plunged into the crumpled strip of country I have mentioned. The hills, though very low, shut off all view, and so intricate were their windings that one had the impression of having wandered by mistake into the scenery of some *montagnes russes.* In half an hour's time our bearings were completely lost, and the caravan was tied in knots. Incidentally we came across the Chinese *dhobi*, who had somehow got detached from the Russian caravan, involved in the same predicament as ourselves the day before. We boast of British phlegm, but for imperturbability of character give me a Chinese *dhobi!* He tacked himself on to our caravan without a word or the least manifestation of relief, and quite unaffected by his night out in the desert with every prospect of spending the rest of his days wandering about in this enormous maze.

When at last we found the pass and climbed to the top of the cliff, a splendid view greeted us. The mountains reappeared ahead, and stretching between them and us was a huge wedge of irrigated country contained between two streams which issued from the mountains some distance apart and united in the plain. In the centre rose a large white tomb, the tomb of Saïd Hassan, which gives its name to all the lands around, while far away on the western horizon was a sight to gladden our hearts — a streak of deep green, the first trees we had seen since we left the Karûn six weeks before.

The green patch was the small Turkish town

of Bedrai, and on the following morning we loaded up a mule with tiffin and started from our camp at Saïd Hassan to pay it a visit. As we rode along I became conscious that a subtle change had come to pass; there was something unaccustomed, something vaguely suggestive of the civilised West in our surroundings. I looked round, but it was the same old familiar Persian landscape of sand and stones, with a few tufts of short dry grass and some grey bushes; then suddenly I hit on the secret—we were on a *road*. For weeks and weeks we had ridden haphazard over a pathless waste, guided only by landmarks on the horizon, but now our horses were following a distinct track which wound visibly for miles ahead, a yellow ribbon across the darker brown of the desert. Soon we began to meet groups of peasants driving tiny pack-cows loaded with trusses of green fodder in little saddle-bags. Then the details of Bedrai itself became visible—an oasis of date groves and orchards, perhaps a mile in diameter, surrounded by a stout mud wall with an entrance here and there, and containing the town itself like a kernel in the centre. We entered by one of the openings in the wall and found ourselves in Paradise.

It came almost as a physical shock, the abrupt passing from the arid, sun-scorched desert which had filled our whole horizon for so many weeks past, into this new world of soft green foliage, fruit-blossom, and running water—almost like plunging on a broiling summer's day into a deep clear pool. We rode at first along a narrow path between mud walls, crossing and recrossing a swift-flowing stream, whose waters were dyed like blood by the soil of the "Red Mountains," whence it came. Masses of may and pomegranate blossom surged over the wall and hung above the path, while a sea of tree-tops of every shade of green stretched beyond. Date-palms reared their serene crests high above into a world of their own, like proud folk disowning any connection with their lowly neighbours around, and a dying frond here and there catching the sun struck a note of dull gold against the surrounding green. Now and again a blue jay or a kingfisher flashed across the path, and the air was full of the noise of cooing doves and the sweet scent of orange blossom. Along the walls at intervals there were heavy timber doors leading into the orchards, each one fitted with a little peep-hole just large enough to allow a tantalising glimpse of the view beyond where streamlets sparkled through the tall, luxuriant grass. One felt like Alice peeping through the keyhole of the garden door and longed for her magic bottle!

Gradually the orchards gave way to houses, and the path resolved itself into a typical Turkish bazaar, very narrow, with houses here and there built across overhead so that we had to duck right down on to our horses' necks to pass underneath. We had planned to picnic among the fruit-trees, so our *mirza* was sent to negotiate the loan of a garden while we despatched an accumulation of telegrams. Through Bedrai runs a small river called by the same name as the place, and formed by the junction of the two streams, the Gavi and the Gundjiuntchem, which enclose, as I have mentioned before, the lands of Saïd Hassan. Our way to the garden led us along a cliff overhanging this river, and there we came on a group of two men and a boy busily engaged in an occupation which puzzled us exceedingly. They had between them two small baskets, one full of worms, the other of short sticks of some black substance. Number one would select a fat worm, hand it to number two, who spitted it with a bit of stick, when it was passed on to the boy, who threw it carelessly into the river below. After several ingenious theories had been expounded as to what it all meant, we asked the men themselves and learnt the true explanation. They were fishermen (save the mark!), and the bits of stick they embedded in the worms were poisoned. The unhappy fish who swallowed the bait died, it appeared, on the spot, and their corpses were retrieved by accomplices with shrimping nets posted farther down the stream. We forebore to moot the delicate questions of who eventually ate the fish!

We found our garden, and passing through a deep gateway in the wall and a porch beyond, came out into a wilderness of fruit-trees, in the midst of which, on a grassy space beside the stream, our host had spread carpets and quilts for our reception. Here, stretched at our ease, with the sunlight dappling our horses' backs as they grazed around us in the lush grass, the blue smoke of our servants' fire curling upwards through the branches, and our own samovar gurgling softly to itself near by, we tasted to its fullest depth the perfect *kaif*, that *dolce far niente* of the East. Into this haven of bliss broke in a vendor of antiques. Experience should have taught me that the one representative of the human race from whose presence one is never safe is the "antica"-seller; yet, I confess, to meet one in Bedrai, which averages perhaps ten visitors from Europe in a century, was a shock. However, there he was with the regular appurtenances of his trade, a number of little objects knotted up in a large pocket-handkerchief. It transpired later that we were near an ancient site called by the Arabs Akr, where they grub up these remains in search for treasure. His collection consisted of a few coins, some sherds covered with hieroglyphics, the head of a little Grecian statuette, a glazed earthenware figure of a rider and horse very roughly shaped (which our expert pronounced to be Elamite), and a polished block of black stone with a swan's head and neck very gracefully carved in high relief. After the usual haggle, in the course of which our friend's original demand abated about 500 per cent, we bought the lot. Thereupon, seeing that business was brisk, he confided that he had at home a unique "antica" of exceptional value, and should he go and fetch it? We assented and waited, full of curiosity to see what his treasure would prove to be. Presently he returned with a fair-sized bundle which he carefully unwrapped, and displayed to our astonished view a dear old Staffordshire china spaniel! He might have come straight from the chimneypiece of some old-fashioned cottager at home, so sleek and smug he looked with hardly a chip off his glossy porcelain coat. His owner put on him a prohibitive price, but I would have given much to know that spaniel's history and all the vicissitudes he had passed through between the potteries and remote Bedrai.

An hour's ride from Bedrai lies the village of Zorbatieh, where we camped for a couple of days. Though surrounded by gardens like the former place, it is far less picturesque, and our stay there was memorable only for the poultry with which we stocked our commissariat. They were a breed of white birds, with wings and tails of bright cerulean blue. The effect was startling, and gave them at first sight the appearance rather of birds of paradise than of barn-door fowls. It was only under close examination that their exotic plumage turned out to be not the work of nature but of a local artist with a pot of paint. Whether their adornment was a compliment to us, or whether it is the fashion of all the hens of Zorbatieh, I was not able to discover. They journeyed with us for several days in a crate fastened on a mule's back, their number dwindling day by day with the same tragic regularity as those of the ten little nigger boys of nursery fame.

Leaving Zorbatieh, we started on the last stage of our journey to Mendeli. It was a three days' march, marked by a couple of unwelcome experiences. On the first day we met with our first and, *alhamdu'llilah*, our last bad sand-storm. By great good luck the wind was from behind, but it blew with so much force that the grains penetrated to our very skins, as well as into the innermost recesses of our luggage, finding its way, as only desert sand knows how to, between the leaves of one's books and into every seam of one's garments, to be finally got rid of only after many days. With the fall of night the storm turned to a tempest, and a terrific thunderstorm, which had been brewing all day among the mountains, burst upon us, straining our canvas to the last stitch. Our tents nobly withstood the onslaught, and only one ridge-pole broke, the pyjamaed inmate of the tent getting a thorough drenching before he could extricate himself from the ruin and find refuge with his neighbour from the fury of the elements.

The next day every ditch and rivulet was in full flood, and the caravan splashed its way through leagues of mud and water. After a while we came to a stream with a strip of bog on either bank. The mules in trying to cross it sank to their bellies, and several had to be unloaded midway before they could flounder through. One poor animal, finding himself stuck beyond hope of recovery, gave up the struggle and quietly lay down on his side in the middle of the stream, pack and all, with the air of one resigned to die. The united efforts of half a dozen *charvadars* eventually hoisted him on to his feet again and hauled mule and load to the farther bank. There, amid universal consternation, it was found that the submerged load contained the priceless original of the "Carte Identique," together with the entire collection of new maps which our survey officers had made up to date. Five minutes later you might have mistaken the desert for the backyard of a laundry, every bush in the neighbourhood being spread with a map put out to dry. Fortunately very little water had found its way into the tin cases in which the maps were packed, and the harm done was negligible.

From our last camp short of Mendeli I rode on ahead to find a camping-ground outside the town. Having selected a suitable place, and being hot and tired from my ride, I made for the friendly shade of a group of palms clustering round a mill. It was an inviting spot on such a scorchingly hot day. A fern-grown aqueduct carried down a volume of clear snow water to feed the mill, the wheel revolving in a vertical tunnel down which the water plunged with a continuous roar. The buildings were enclosed by a wall pierced by a deep pathway, within the

1. The British Commission. Back row, left to right: G.E.H., Col. Ryder, Capt. Wilson. Seated: Capt. Brooke with 'Ben', Major Cowie, Mr Wratislaw, Capt. Pierpoint with 'Sheila'

2. Pillar No. 1 surmounted by Basri Bet, my Turkish colleague

3. The colonel and the cyclometer

4. The British 'corps technique' at work

5. Mohammerah. On the bank of the Karun

6. The Ashar, the port of Basra

7. The palace of the Sheikh of Mohammerah, on the Shatt-el-Arab with his private theatre on the left

8. The State lunch given us by the Sheikh of Mohammerah. There were about 250 dishes on the 'Table', including a whole roast sheep

9. Loading up. Mules and a Yabu

10. Ibrahim and Mullah Hassan, two of our muleteers

11. Ten sand grouse at one volley – sitting! Hardly sporting but we wanted them for the pot

12. Crossing the Kerkha

13. The Marsh Arabs' country

14. One of their reed huts

15. The marsh-village of Bisaitin

16. Our guide Ibrahim (aged 80)

17. The Duerij, when we reached it, was only 100 feet across. After the British Commission crossed, it rained for 48 hours on end and the river became a hurtling flood

18. ... impassable for our colleagues on the other bank, who had to wait till ...

19. . . . the waters began to subside . . .

20. . . . when they got across with great difficulty

21. The desert is broken by a number of small rivers flowing westwards from the mountains to the Tigris

22. The Tyb, one of the larger of these rivers

23. Fording the Tyb

24. Fording the Tyb

25. A mule-load is about 240 lbs in addition to the weight of the big bells on either side

26. The arrival of the mails – they were carried up the Tigris to points opposite our line of march and brought thence by a 'kassid' escorted by an Indian sowar

27. The Wali's Vizir

28. The Wali's horse, with gold and silver bridle and leopard skin saddle

29. The Wali of Pusht-i-kuh coming to call in state, preceded by his foot-guard . . .

30. . . . and followed by his horsemen

31. When he left . . .

32. . . . we returned the call

33. The Lost Caravan. The 'brown paper' hills

34. Confidence

35. Doubt

36. Confusion

37. The return to civilization. Bedrai, on the Turkish side of the frontier

38. Mendeli and its gardens

39. Bagdad. The bridge of boats

40. The ferry. 'Gufas' are the characteristic craft of Bagdad and are as ancient as Nebuchadnezzar

41. The Bagdad-Kermanshah road. Our 'bus' halted in the courtyard of the Khan at Kizil Robat during the 100 mile drive from Bagdad to Khanikin

42. Kasr-i-Sherin. An ancient capital of the Sassanian kings, called after the beautiful queen Sherin ('the Sweet')

43. Kurdistan. Kurds of Halebja watching

44. A typical Kurdish village built in tiers up the mountain-side

45. Kurdistan. Our caravan crossing the river Sirvan

46. The caravan passing along the foot of Mount Bamu

47. A street in Tavila . . .

48. Gulambar, the village of 'Amber Rose' is as pretty as its name

49. Nakishbendi Dervishes

50. A well armed group

51. The chief of the Mangur tribe at Vezneh

52. Kurdish full dress

53. A Kurdish village

54. Kurdish landscape

55. A stronghold of the old Kurdish chiefs in a deep ravine, abandoned by their modern descendant

56. Ushnu. Musa Khan, son of Mansur ul Mamalik, Chief of the Zerzaw. He was only 13 but was married and a confirmed gambler

57. Cossacks of the Ushnu garrison

58. An Anglo-Russian 'entente'

59. The grain market at Urmia

60. Finis. One of the last of the pillars

61. Finis. Rounded up. Simko and his men with six of the tribesmen who attacked us. Two of the latter were afterwards hanged at Tabriz

shade of which sat the miller with his family—that is to say, his wife and two brown little boys, who looked delightfully quaint in their high Persian hats and blue shirts open right down to their fat little tummies. I was standing watching the bare-legged bakers come riding up on their donkeys to fetch away sacks of flour when the jovial miller spied me. He at once brought out a carpet and led me to a strip of grass by the side of a pond among the palms, where he spread the carpet and begged me to sit down and rest. His act was typical of the innate courtesy of the Persian, however lowly born, and his first remark was an example of the pretty turn of compliment which comes so naturally to their lips. I had admired the smooth green turf and the clear pool at our feet. "The praise to Allah!" he replied; "the grass is greenest and the water clearest wherever your presence brings honour." Our conversation turned to the question of the frontier, in which my miller took a personal interest, being a Persian himself, but living actually on Turkish soil. He asked how the line would run past Mendeli. I sketched it for him, and asked if he approved. "*Che arz kunam?*" was all he said. "What petition shall I make? It is all in the hands of Allah."

After all, I reflected, what higher compliment could the worthy fellow pay than to identify the Commission thus with Divine Providence!

Mendeli is the chief town of a *kaza*, and boasts

a *kaimakam;* it also has a fine roofed-in bazaar and a remarkable system of small canals, which wind in and out, and even tunnel underneath the houses, in the most unexpected fashion. In spring-time these canals are full of clear, swift-flowing water, and lend a particular charm to the place; but in summer, when they dry up, they must become pestilential in the last degree. We called on the *kaimakam*, who received us very affably, and read aloud to us the latest news of the Ulster crisis from his copy of the local Turkish newspaper.

Bagdad lies seventy miles west by a little south of Mendeli, the nearest point of the frontier. By the time we had reached this latter place the temperature in the plains was rapidly increasing, being already 100° in the shade. There remained still eighty very difficult miles of frontier to demarcate before the hills were reached, and in consequence the Commission found itself involved in a race against time.

Very reluctantly, therefore, I saw my hopes of a lifetime fade as the prospect of being able to see Bagdad grew more and more remote. As chance would have it, however, the proverbial wind which always brings good to *somebody* took this occasion to blow, and wafted me all unexpectedly to the desired port. The heat and the fatigues of the long desert march had affected the health of our chief, Mr Wratislaw, and he was ordered by the doctor to take an absolute rest and recoup among more civilised surroundings. I was to go with him to Bagdad, and so it came about that, accompanied by four sowars and a caravan of fifteen mules, we left Mendeli on the morning of April 15 and set our faces towards the Tigris.

Except for the storm of a few days before, the weather had been "set fair" for some weeks past, and a few white clouds hovering over the horizon looked unfamiliar. By ten o'clock they had grown bigger and blacker, and we saw that we were in for one of the fierce thunderstorms which sweep along the edge of the plain at this season of the year. There was not a scrap of shelter from the tearing wind, so we ate our tiffin as best we could behind a screen formed by a blanket rigged up against a couple of lances stuck in the ground, and then rode on through the pouring rain. We had dried in our saddles when, after eight hours' riding across the naked brown plain, we saw ahead of us the welcome dark patch which betokens an oasis village. But there's "many a slip" for the traveller in these pathless regions, and just as we were congratulating ourselves on reaching the end of the day's march, we came to the brink of a muddy lake, a mile or more wide, cutting us off from the village and stretching indefinitely in either direction. The flood apparently covered the desert for miles around, so we chanced our luck and rode into it girth-deep. By Allah's favour the caravan struggled safely through with the exception of one mule, who lost his footing and collapsed into three feet of water and mud amidst the wreckage of our kitchen.

The village we had come to was Balad Ruz, one of the most charming I have ever seen. As we rode through next morning in the freshness of dawn, it looked so pretty as to seem almost unreal. The path wound alongside a willow-shaded stream, the innocent-looking author of our kitchen's ruin! The young sunlight fell in splashes of gold on the water and the high mud walls which hemmed in the path on either side, and over which fell great scarlet sprays of pomegranate, brushing our shoulders as we rode along.

Nearer to the village itself monumental piles of dry sticks crowned the walls, from the top of which sleepy storks gazed disapprovingly down at us, one here and there laying his head flat along his back and clacking his beak like castanets in protest against such early risers. (This trick, by the way, has earned for them a quaint name among the Arabs, "Hajji Lak-Lak," *hajji* being the title given to a pilgrim back from Mecca, whither the storks are supposed to disappear at their annual migration.)

An hour later we had left this pretty village far behind, and were once more swallowed up by the vast empty plain. That evening brought us to Bakuba, a large village on the Diala, which is an important tributary of the Tigris. We rode through a long covered bazaar, very gay with stacks of oranges and piles of bright-coloured stuffs in the booths on each side, and on over a crazy pontoon-bridge, to our camping-ground on the farther bank.

The next day saw us posting the remaining thirty miles into Bagdad in a four-horse landau, the local form of post-chaise. We drove along the Bagdad-Kermanshah road, but the ill-defined track of wheel- and hoof-marks over which we jolted and swayed was hard to recognise as the chief trade-route between Turkey and Persia, and one of the greatest pilgrim-roads of Asia. Along it flock the many thousands of devout Shias who make pilgrimage to Kerbela and Nejf, the *mesheds* or "places of martyrdom" of the Caliph Ali and of his son Hussein, as holy in the eyes of the Shias as Mecca itself. They lie on the farther side of Bagdad, not far from the ruins of Babylon. There are other places of pilgrimage at Bagdad itself, including the tombs of the seventh and ninth *Imams*.

Bumping along this execrable road worn by the feet of the faithful, we met and passed pilgrims of every degree. First would come some rich man's train. He is doing the pilgrimage *en grand seigneur* with a team of pack-mules and *kajavehs* for himself and family. These are a sort of closed pannier slung one on either side of a horse. They rock atrociously, and the occupant, who squats on the floor, must experience all the sensations of a cross-Channel journey. A muffled twittering issues from one as it passes, betraying the *harîm* within; another, with the curtains thrown back, accommodates the solemn and bulky figure of the great man himself. Next come humbler folk on foot with donkeys to carry their belongings; and last of all, the poorest pilgrims who cannot even afford this luxury, but carry their little all in bundles on their backs. One old man we stopped and asked where he came from. He was a tired-looking ancient, and could only hobble with the help of a stick, but he cheerfully answered the name of a small town on the farther side of Teheran! Poor old fellow, he must have been weeks on the road, and he was quite alone; but we left him contentedly munching a morsel of stale bread and garlic which our driver, in a fit of generosity, had spared out of the remains of his breakfast.

It struck me as I watched them pass what a loss to Christendom it must have been when the old pilgrimage habit died out. Your Moslem *hajji* loads up a donkey, sets off from home, and tramps for months across whole countries or even continents. (I have read a book of travel in Central Africa whose author came across a party of pilgrims somewhere near Lake Chad on their way to Mecca.) On his way he falls in with other *hajjis* of a dozen different races, all making for the same goal and united by a common bond of fellowship. He reaches the sacred shrine, performs the established rites, and returns home with a stock of yarns to last a lifetime and a comfortable feeling inside that he has booked his seat in Paradise. Think what an experience it would be for a British labourer or artisan!

When we were still two hours distant from Bagdad the mirage ahead began slowly, fantastically, to resolve itself into a vision of domes and minarets. It was a strange ethereal city which gradually took shape before our eyes, for the haze obscured the lower half of each building, leaving its upper portion floating in the sky. Presently, farther to the north, a broadening dark streak betrayed the course of the Tigris with its double belt of palms. Then two points of light beyond began to flash in the sun, which we knew for the Imams' tombs of Kazimein with their domes plated with solid gold. It was surely such a vision as the Dreamer on Patmos must have seen when he described the jewelled city of the Apocalypse.

But the vision was fleeting, and disillusion followed close at heel. The dream-city soon became a solid thing of brick and mortar, and the nearer we came to it the less ideal it appeared. Bagdad is, *par excellence*, a river-town. She owes everything to the Tigris, and honourably repays the debt by reserving all her charms for the river front. To the desert she turns her back, and a remarkably ugly, sordid back it is. Entering thus, so to speak, by the back-door, our first impressions were anything but favourable. The carriage drove in through the *débris* of the old town walls, threaded its way between a wilderness of pariah-haunted cemeteries and dust-heaps, relieved by a solitary fine old tiled minaret derelict amidst the surrounding waste, and plunged into a region of featureless back streets. From these depressing scenes we emerged presently into the main street, and ended our journey a minute later at the gates of the British Residency.

CHAPTER VIII.

DAR-EL-KHALIFEH.

Certain towns of Turkey possess titles which constitute a sort of official surname. Constantinople, to take an example, which in ordinary conversation is plain *Istamboul*, assumes in documents of State —and even nowadays in the columns of newspapers—the more imposing guise of *Dar-el-Saadet*, "The Place of Felicity." Bagdad, in the same way, boasts the proud title of *Dar-el-Khalifeh*, "The House of the Caliphate" (or, as we know it better, "The City of the Caliphs"), and so keeps green the memory of its golden age. Of all the cities where in different ages the Commander of the Faithful has held his court—Damascus, Cordova, Cairo, and in these latter days Stamboul—none is closer linked with the triumphs of Islam, yet none has fewer relics to show of its glorious past. Floods, fires, destroying armies have done their work too well, and obliterated almost the last trace of what in the ninth century was perhaps the most splendid city of the East.

It may be interesting to outline briefly Bagdad's

history. There is some evidence that a town existed on the present site in the days of Nebuchadnezzar. A fragment of wall on the river's bank was discovered last century built of mud bricks, each one stamped, like those of Babylon, with the name of the herbivorous king. It is very striking, by the way, to a traveller in Mesopotamia to see how permanent a building material mud is—far more so than stone. The reason is twofold: firstly, mud architecture must, by the nature of things, be built exceedingly solid; and secondly, mud walls are hardly worth pulling down and carting away, whereas stone can be always used again, and a stone-built ruin is regarded purely and simply as a convenient quarry. If Ctesiphon had been built in stone, I greatly doubt if a single stone of the colossal arch would be standing to-day.

But to return to our subject. It was the Abbaside Caliph Mansour who, in search of a capital worthy of the head of Islam, then at the zenith of its martial glory, chose Bagdad for the site. This was in 762 A.D., a bare hundred years after Mohammad's death. Half a century later Bagdad had become, if not the hub of the universe, at least the centre of gravitation of the Islamic world, gathering to itself the highest in art and science of the age. The Caliph's court, the most splendid in the East, rose to the highest pitch of magnificence under Haroun-er-Rashid, immortalised for ever by the tales of the 'Thousand and One Nights.'

The inevitable decadence followed, and the Caliphs became mere creatures of luxury, and finally puppets in the hands of their own servants. In 1277 Hulagu Khan swept down on Bagdad with his Mongol hordes and destroyed the last of its Caliphs. The subsequent history of the Caliphate, how it was transferred to Cairo, and thence in the sixteenth century to the reigning dynasty of the Sultans of Turkey, does not concern our subject. The ubiquitous Tamerlain found time to visit and sack Bagdad. Ottoman domination began there in the reign of Suleyman the Magnificent, the contemporary of Queen Elizabeth, who, with his successor Selim I., extended the boundaries of the Turkish Empire to their farthest limits. In the seventeenth century, when that empire had begun to wane and Persia was a formidable rival, Bagdad fell to the armies of Shah Abbas. It did not remain long, however, in Persian possession, but was recovered in 1638 by Sultan Murad the Fourth.

Since then there is little to record. Though retaining always a considerable degree of importance by its past, its situation on the Tigris, and its position as capital of the great province of Irak, the town sank more or less into provincialism, and became, from the point of view of the Porte, chiefly a convenient place for exiling obstreperous Valis. Later, its importance was, of course, enormously enhanced when the Bagdad Railway began to take concrete shape, and at the time of the outbreak of war was in a fair way to regaining much of its old pre-eminence, as soon as this immense scheme should have reached completion.

Extremely few authentic relics of the days of the Caliphate remain, as I have already said, at Bagdad, and the traveller who arrives there, primed with memories of the Arabian Nights and hoping to find himself again among their scenes, is doomed to bitter disappointment. Happily for myself I had been forewarned, so I arrived prepared to take Bagdad simply for what it is to-day, a very interesting and picturesque Turkish town.

The first feature of Bagdad that struck me was the number and colossal impudence of its swallows. The house was full of them; they were building nests in every available corner, and at dinner-time they flew in circles round the table. When I went to my bathroom on the first morning I found a young pair setting up house, so I drove them out and closed the shutters. On coming back to my bedroom I found them hard at work there, and again I expelled them. They merely transferred their energies to my sitting-room, which apartment having four windows and two doors, I realised that further action was hopeless, and resigned myself like the other inhabitants of the place to a policy of contemptuous indifference.

All good travellers to Bagdad go first to see the bridge of boats, so thither we dutifully set out on the morning of our arrival. But there had been an exceptionally heavy spate on the river a few days before, the water rising nearly 20 feet in as many hours, and the bridge was swung back along either bank out of harm's way. The scene on the river, however, was inspiring. A steam-tug with loaded *sefinehs* lashed on either side came tearing down-stream at a good fifteen miles an hour; ugly iron lighters, which had been towed up from Basra full of rails and boilers for the Bagdad Railway, were discharging their cargo on to the western bank with a fiendish clatter and din, and dozens of *gufas*, the characteristic small craft of Bagdad, were plying across the river. The *gufa* is the most archaic type of boat in the world. It is simply a big round basket of wicker-work—the largest holding twenty passengers and more—" pitched within and without with pitch " like Noah's Ark. It spins helplessly round and round in the stream, apparently quite unresponsive to the efforts of the rowers, and you can visualise it exactly if your childhood memories are fresh enough to recall the picture in the nursery rhyme book of the " Three Wise Men of Gothan who went to sea in a tub."

The bridge being in normal times about as crowded as London Bridge at noon, the dislocation of traffic caused by the flood was considerable, and the *gufas* were doing a roaring trade. A small fleet of them lay in a basin near the bridge-head. Each passenger as he arrived at the landing-steps bundled into the nearest, and when there was not space left for a mouse the *gufa* cast off. It was towed some hundred yards up-stream by a man with a rope walking along a ledge built out for the purpose from the river-walls of the houses lining the bank, and then launched on its voyage towards the farther bank, which it reached, after strenuous efforts by the rowers, at a point a quarter of a mile or so down-stream. Occasionally a merchant turned up, followed by a string of porters with bales of merchandise. Then an extra large *gufa* would be requisitioned, the bales piled inside till they formed a pyramid eight or nine feet high, their owner would scramble up and perch himself perilously atop, the rowers squeeze into any crannies they could find below, and the whole marvellous erection sail off.

Parallel to the river, and extending through half the length of the town, run the bazaars, connected by short cross-streets with the river-side wharves. They are roofed in almost from end to end—in parts with a high-pitched timber roof, and in parts with what the Turks call "Aleppo" vaulting—a series of traverse arches with the spaces between filled by small brick domes. Through apertures piercing the roof at regular intervals the sun streams into the dark interior and forms shafts of light all a-whirl with a grey-blue haze of cigarette smoke. The effect of these vistas of bright columns against a dim background is delightful. Two features which I always welcome in an Eastern bazaar were, however, conspicuously absent—the all-pervading smell of saffron and the rhythmic thud-click-click-thud of the spice-grinder's pestle and mortar; but there was compensation for the loss in the brightness of the colour scheme. There is a glad note in the architecture which is lacking in the more austere styles of Cairo and

Stamboul — the result, no doubt, of Persian influence. The gateways leading to mosques are many of them covered with green, blue, and white tiles, with here and there a splash of red, set in minute and intricate designs. Dainty little archways surprise you at odd corners, and now and then you catch a glimpse of a tile-clad minaret or dome encircled with a broad frieze of graceful Kufic script. At one place in the main bazaar there is a saint's tomb hidden behind a delicately painted façade of pink roses. The roses are all worn away at the level of a woman's lips; I suppose the saint has some special blessing to bestow on his fair devotees.

Bagdad must have as heterogeneous a population as any city of its size in the world, and in a walk through its bazaars you meet types of a score of different races. Predominant, of course, are the native Arabs. They retain, even in the degeneracy of town life, the dress of the desert— the long brown cloak, and for head-dress the *kuffiyeh*, held in place by a halo-shaped *agal* of twisted camel's hair. But the former is rarely now the tasselled red-and-yellow silk handkerchiefs of old days, but usually a piece of cheap and ugly "Manchester goods" hardly distinguishable from a housemaid's duster. Then there are the Persians, merchants for the most part—fat, placid gentlemen of the type you see portrayed in old Persian miniatures. They have a certain air of unctuous dignity as they sit in their narrow booths, tight-packed among their wares and toying with the beads of an amber rosary.

A chance sign-board above some humble shop catches your eye: the name is one that figures high in the roll of titled plutocracy nearer home. A successful race these descendants of the Children of the Captivity who still haunt the banks of the "Waters of Babylon," but send their off-shoots to control the financial destinies of Europe!

Presently you meet a string of sturdy ruffians paddling along at a jog-trot with Atlantean burdens on their backs. Dressed in short kilts, with red handkerchiefs twisted round their heads, and murderous-looking knives stuck in their belts, they look like a gang of stage pirates. These formidable fellows are Kurdish *shayyâls* or porters—superb specimens of humanity, with a development of shoulder and leg worthy of a Rodin statue. They were the first of their race I saw, and did not at all accord with my preconceived idea of what a Kurd should be. Instead of the dark, brutal faces with which my fancy had endowed them, many of them had blue eyes, fair hair, and refined features, often startlingly Anglo-Saxon in type, and a remarkable contrast to the expressionless Oriental faces of the crowd around.

There is a large native Christian element, lineal descendants, some of them, of the ancient Chaldeans, and adherents now of one or other of the many Churches which flourish so abundantly on the soil of Mesopotamia — Catholics, Jacobites, Syrians, Greeks, and Nestorians. Besides the ruling Turks — chiefly Government officials—there are Armenians and representatives of I know not how many more races, including a considerable body of Indians, either merchants settled in trade or pilgrims to the famous Sunni shrine of Abd'ul Kader.

The mention of these last recalls a rather interesting fact connected with the duties of the British Consul-General at Bagdad. It seems that in the old days the Honourable East India Company once found themselves badly in need of cash. The King of Oudh, hearing of this, offered to lend them a few scores of lakhs of rupees, but on a special condition—namely, that the capital sum was never to be repaid, but that the Company should undertake to administer the interest in perpetuity for the benefit of certain religious charities mentioned in his will. These terms the Company accepted, and they and their successors—the Government of India—have loyally fulfilled them ever since. Several thousands of pounds annually were bequeathed in the royal testament to the deserving poor at Kerbela and Nejf, and this the British Consul-General has to distribute. Twice a year he makes the arduous journey to the shrines, where he presides at the sittings of Relief Committees, and takes such steps as he can to limit the proportion of the sum which adheres to the palms of the local *mujtehids.* Returning from this errand while we were guests in his house, he told how he had been importuned by an impecunious Indian who wanted a job as a clerk. Inquiries as to the applicant's identity and antecedents disclosed the fact that he was the great-grandson of the last King of Delhi! *Sic transit* . . .

One day we paid the usual formal visit to the Vali. He was a fine-looking old soldier—one of the few Turkish generals who came through the Balkan war with credit. His talk was of schemes for the development of his *vilayet*—an airy project to establish a motor-bus service across the desert from Aleppo, and the more concrete irrigation scheme which Sir William Willcocks had worked out on the spot some years before. Ever since the days of that great reformer, Midhat Pasha, Bagdad has been blessed with "progressive" Valis; but progress, as conceived by the Ottoman mind, sometimes leads to queer results. This was the case under a recent Vali. The main street of the town was, as it still is, extraordinarily narrow, and His Excellency had the commendable idea of having it widened. His method was simple. A straight line was drawn to mark the new limits of the street, and whatever projected beyond that line was shorn off till it reached it. The consequence was a row of what looked like open doll's houses, with the whole of their internal economy displayed to the public view. An admirable system of ventilation was, I believe, the only compensation that the unfortunate owners got.

We had been told by Javid Pasha that a review of troops was to be held on the following day to celebrate the anniversary of the Sultan's accession, so duly went out to the racecourse behind the town to witness the spectacle. There were very few troops available, and rather pathetic efforts had been made to fill up numbers with cadets, boys from the Industrial School, and suchlike—the "make-weights" even including a cart carrying a crate of carrier-pigeons (which our Christian dragoman maliciously suggested had been bought in the bazaars that morning). After the march past of the troops an amusing scene took place. The guilds of Arab craftsmen had been ordered to parade with the rest. They duly collected around their respective banners, and Turkish officers arrived to range them into line. But the craftsmen, having no love for the Turk and little respect for the Vali, had their own programme, and began to work up for a war-dance. It started by some one intoning a monotonous, ever-repeated chant. One by one the others picked it up, gradually increasing the time and marking the cadence by leaping into the air on each leg alternatively, waving meanwhile whatever weapons they had in their hands above their heads. Before long the chanting had merged into one deafening roar, and there was a raging mob of lunatics, brandishing knives, rifles, and sticks, seething round each of the banners. A group began to sway backwards and forwards, then, without warning, leaderless, and as if possessed by an irresistible common impulse, broke away and came surging wildly across the ropes right into the middle of the spectators. It was a really alarming sight, and had we not been accustomed to the innocent eccentricities of an Arab war-dance (in peace time), it would have needed some nerve to save us from taking to our heels in flight. The efforts of the Turkish officers to control their charges and bring them back into line were pitiful, and the whole scene resembled what I should think a Cumberland sheep-dog competition might look like with exceptionally wild sheep and quite untrained collies. For a moment the entire yelling mass would come dashing forward in the most promising style, when suddenly, and for no perceptible reason, it would abruptly right about turn and go rushing back again. At last—by pure good luck, I fancy—the collie-officers got them all heading straight for the Vali, and in one wild rush they went past the saluting-point. One hoped His Excellency felt gratified!

I have mentioned the distant glimpse we had, on nearing Bagdad, of the golden domes of Kazimein. Here at least was something which smacked of the Arabian Nights, and I was full of curiosity to get a nearer view. But the tombs of the

Imams are so zealously guarded by the fanatical Shias who live round the shrine that an infidel may not even approach within sight of the gates. Only by strategy could one get near enough to see, so when I was invited by old Mirza Daoud to take tea at his house at Kazimein, I accepted with alacrity. Mirza Daoud was Captain Wilson's confidential servant, and an old Persian of great character. He had travelled far and wide with his master, and, like the traditional British tar, had "a wife in every port." The fairest of them dwelt at Kazimein, and now after an absence of five years he had returned for a week of connubial bliss, and incidentally—as he naïvely explained—to investigate a little matter of several sums of money which he had remitted to her for repairs, but which the fair lady, it seemed, had appropriated for her personal adornment, leaving the house to go to rack and ruin.

On the appointed day I set out for Kazimein. The first half of the journey was by motor-boat up the river, past delightful old houses and gardens, wharves and warehouses, *serais* and mosques, whose blue-and-white minarets leaned like the tower of Pisa, coffee-shops built far out above the water to catch the breeze, and creeks between, where high-pooped *seffinehs*, with prodigious outrigged rudders and tillers as large as a fair-sized tree, discharged their cargoes of cotton and brushwood. A prehistoric horse-tram took me the rest of the way, and put me down in the filthiest and most closely-packed Eastern town I ever had the misfortune to visit. My host's house I found to be a quaint little place built round a courtyard, barely large enough to contain the solitary palm-tree which adorned it. A flight of stairs—about which the less said the better—led up to the guest-room, arranged with an open arcade along the whole of one side to allow the cool breeze to blow in from the north. After absorbing several cups of sweet Persian tea out of a cup which adjured one to "Love the Giver" (a fine testimony to the penetration of British trade), and listening to the old fellow's yarns of his adventures with his Captain Sahib in the wilds of Luristan, I proposed an excursion to see the domes. Mirza Daoud had a friend who kept a caravanserai in the neighbourhood of the tomb, so to him we went and asked permission to go on to his roof. The friend proved complaisant, and after one moment of awful suspense, when, having deprived me, as a measure of precaution, of my *ferenghi* hat, he appeared to be on the point of clapping his own turban on my defenceless pate, we were allowed to climb up. There, a couple of hundred yards away, we saw the famous domes, flanked by their attendant minarets, glowing in the light of the setting sun. At that distance the surface was not easily distinguishable without the aid of glasses, which it would have been rash to use in such an exposed position; but it seemed to me that the whole of either dome and the upper quarter or so of each minaret were plated with scales of the precious metal laid on like tiles. The effect, I must confess, made more appeal to the imagination than to the eye, and my material Western mind was rather engaged in a sordid speculation as to the value of the gold than in any attempts to judge of its artistic effect. What the figure may be I have no notion, but it must be something prodigious.

As we threaded our way home through the narrow streets, I asked my companion who the author of such munificence had been; he told me it was some ancient king of Persia who had made *ziaret* to the tomb. Just then we passed the city gates, through which trickled a thin stream of pilgrims. Here they were, after weeks or months of tramping, arriving, as it were, at the very gates of Rome, and I scanned their faces for some trace of the emotions which a man might show at such a crisis; but none of the religious fervour that moved the old Shah to his lavish deed was visible in them, no flash of eye or quickening of step, nothing in their bearing but a tired indifference.

One last impression of Bagdad remains firmly fixed in my memory. We were watching the sunset from the terrace of the Residency on the evening of our departure. A pearly-blue haze hung down-stream, broken only by the yellow mass of a big Turkish palace built on the bank a little further down. Up-stream the contrast was extraordinary. An arch of liquid orange light curved above the horizon, its zenith directly above the river. Tawny reflections fell on the river's surface, and battling with the shadows among the ripples, turned it all into a silken glory of shot blue and gold. Palms on the banks and masts of ships stood in sharp relief against the sunset, as hard as the landscapes in bronze on Ghilberti's doors in Florence. There was hardly a movement or a sound to break the peace. Suddenly a German tug came round the bend, loaded with iron rails for the "Berlin-Bagdad" Railway, forged up the reach, steered for the opposite bank, and a moment later shattered the air with the desecrating clatter and roar of her anchor-chains. How clear an omen, had we but known it!

That night we left Bagdad on the Khanikin mail on our way back to the frontier.

CHAPTER IX.

MESOPOTAMIA IN RETROSPECT.

BEFORE following our travels into Kurdistan, I propose for the space of one chapter to arrest my narrative, and look back at the immense plains of Lower Mesopotamia from a more general point of view than that of the mere traveller. At the time at which I am writing even the hardest political prophet reserves his prognostications concerning the future of this region. But of this one may be fairly sure that, whatever happens, our connection with it must be strengthened rather than diminished now that British soldiers have fought and died on its soil.

The potentialities of this huge treasure-house of Nature, sealed through a thousand years of ignorance and misrule, and waiting only the "open sesame" of the modern irrigation engineer to unlock again its portals and supply food for half a continent, has been insisted upon by writers galore, and if any material guarantee were needed of the truth of what they say, the capital sunk in the

Bagdad railway and the Mesopotamian irrigation scheme amply provide it. I will therefore content myself with repeating what was said of it in an official report made many years ago to the British Government, that "it produces all the grains of Europe in abundance, together with rice, maize, sugar, cotton, indigo, mulberry for silk, and every sort of fruit in profusion"; to which list may now be added that substance of incalculable future value, petroleum oil. Innumerable traces of ancient canal systems which scar the whole face of the land must stir the imagination of even the merest layman, and turn his mind to speculating on the changes which will result if stable government and scientific methods are ever reintroduced.

Even with things as they are, however, Lower Mesopotamia is, and has been for a very long time, an important market for European and Indian trade. British merchants were established at the head of the Persian Gulf before the middle of the seventeenth century, not very long after they first attained a footing in India, and the history of their first settlement, and the subsequent development of our trade in the country, form a tale worth telling.

Actually the first commercial relations between England and the countries round the Persian Gulf came about in the most roundabout way imaginable. A society of gentlemen and merchants, formed for the discovery of unknown countries, equipped in 1553 a fleet of three ships, the *Bona*

Esperanza, the *Edward Bonaventure*, and the *Bona Confidentia*, under the command of Sir Hugh Willoughby and Mr Richard Chancellor, to try and find a passage along the north of Siberia to China. After terrible sufferings, in the course of which poor Sir Hugh and the entire companies of two of the ships were frozen to death, the survivors reached Archangel, and Mr Chancellor found his way to Moscow, where, supported by letters from Edward VI., King of England, he obtained privileges which led to the creation of the Russia Company.

Three years later four more ships were sent, and one intrepid Englishman of the name of Jenkinson penetrated as far as Bokhara and Persia, where he is said to have met with merchants from India and Cathay. This journey he accomplished seven times, trading later with Russia and Persia. The only European competitors in the field seem to have been the Venetians, whose cloths, drugs, &c., came overland by Aleppo. The war between Turkey and Persia which was then raging interfered, however, with this circuitous trade route, and by 1581 the enterprise had ceased altogether. Certainly those old sixteenth-century traders well earned their names of "merchant adventurers."

It was another and far more celebrated "Company" which next established a connection between England and Chaldea. In 1614—that is, two years after they had set up their first "factory" in

India—the Honourable East India Company turned their attention to Persia and the Gulf. The origin of this adventure was again rather out of the ordinary. A certain Mr Richard Steele went to Aleppo to recover a debt from a merchant of that city. The merchant—who must have had a wholesome fear of his creditor—fled through Persia to India, pursued by Mr Steele, who, having reached Surat, made a report to the East India Company on the great opportunities awaiting them in trade with Persia. The Company acted on his hint, and there began the great struggle for supremacy in the Gulf with the Portuguese.

A factory was started about 1630 at Gomroon (Bunder Abbas), and in 1639 the Presidency of Surat sent two of their agents to Bussorah, as it was then always called, in order to establish another on Turkish soil out of reach of the oppressive measures (due partly to the influence of their Dutch rivals) to which they were subjected in Persia.

Of the earliest days of the Basra factory no records have survived, but in 1646 Thomas Cogan and W. Weale write from that place to their "worshipful and much-respected friends" at Surat. Trade is not brisk, for "the pepper is all landed, but not a merchant will proffer more than three rupees, which, if we would make sale of it at all, at that price it should go, but they who proffer it will not have more than three or four *maunds* at most, so that perforce we must keep it; we hope for a better time," also, "the blue cloth remains still in house, nor yet any offering to buy." Again in 1661 two of the Company's gentlemen are sent on board the frigate *America*, "now by God's permission bound from Swally Hole unto the port of Bussorah, within the river Euphrates"; they are enjoined to salute the *Bashaw* and ask for a better house than he had hitherto provided for the storage of the cloth, pepper, cassia, lignum, rice, &c., which they were bringing with them as merchandise.

In the January of this year a document had been signed at Adrianople which had the most important results for our trade in Mesopotamia, as well as in every other part of the Ottoman dominions. This document, which consisted of the first Capitulations between the King of England and the Sultan, is conceived in the exalted phraseology of the Sublime Porte, and commences thus:

"By the favour of the Nourisher of all things and mercy and grace of the Merciful, I that am the powerful Lord of Lords of the World, whose name is formidable upon Earth, Giver of all Crowns of the Universe, Sultan Mahomed Khan.

"To the glorious amongst the great Princes of Jesus, reverenced by the high potentates of the people of the Messiah, sole director of the important affairs of the Nazarene nation, Lord of the Limits of Decency and Honor of Greatness and

Fame, Charles the Second, King of England and Scotland, whose end and enterprise may the omnipotent God conclude with bliss and favour with the illumination of his holy will."

It is hard to believe that even Sultan Mahomed's stern eye can have been *quite* innocent of a sly twinkle when he dubbed the Merry Monarch with this most ambiguous title.

These capitulations (the Sultans of those days would not demean themselves to the level of Christian monarchs by concluding *treaties* with them) assured many advantages to English merchants. To name a few, there was to be freedom of travel and freedom for ships to trade, with a fixed import duty of 3 per cent, protection in lawsuits, a stipulation that Englishmen should not be slaves, provision against pirates, and a "most favoured nation" clause.

There is little to tell about the development of the East India Company in Mesopotamia for the next seventy years. Basra remained the only factory, but trade flowed thence northwards and eastwards by the boats to Bagdad and the Aleppo caravans. One of the factors' chief duties was to forward to India the letters and news which arrived by the latter route, and their letters to Bombay are often a quaint jumble containing scraps of such news as the death of the King of England, or the signing of peace with Spain, sandwiched in between their trade accounts. In 1733 there was war between Turkey and Persia over the deposition of "Shaw Thomas," as the then Shah is called in the correspondence. The Persians, who had taken Basra, tried to draw in the Company and secure the help of their ships, but apparently without success. Life must have been precarious in those days at Basra; for besides the frequent attacks on the town by the Arabs of the surrounding desert, it was periodically visited by the plague. The Diary of the Bombay Council for 1738 contains, for instance, the following laconic statement: "There has been a general sickness at Bussorah, by which all the English gentlemen were carried off except one, Mr Sterling, who prudently sealed up the warehouses till some one should arrive to take charge." In return for such risks they drew munificent salaries of from £5 to £30 per annum, but it is to be presumed that their chief profits came through their private trading. There seems, at the same time, to have been occasional perquisites of a sort to go some way towards compensating the factors for their discomforts and exiguous pay. Mr Shaw, for example, whom we find representing the Company at Basra in 1754, reports that "about twenty days past I was surprised with a very singular instance of Solyman Bashaw's (of Bagdad) respect and regard for the English, he having despatched hither a very principal officer of his household purposely to salute me in his name, who arrived after a passage of thirty-five days, and by him sent me a fine Turkish sabre with a very noble horse,[1] richly caparisoned with gilt furniture, with very extraordinary letters of compliments from the Bashaw, his First Minister and Master of Household, expressing the particular satisfaction they received from the regular decent conduct of our nation at all times." Mr Shaw found himself obliged to reward the bearer of these gifts with a fur coat, "being informed that such was the custom observed among Turkish courtiers," but hopes he has his Hon'ble Employer's liberty to charge its cost to their account!

In 1749, after a period of unrest occasioned by the arrival at Basra of a new *Bashaw*, whose first act was to cut off the heads of twelve of the principal "zanysarees," to which the populace replied by seizing ten field-pieces out of the magazine and levelling the *Bashaw's* house with the ground, trade again began to boom, and the Resident is able to write as follows: "On the 28th ultimo sailed the *Prince Edward* for Bombay with a small freight. She has made the greatest sales that ever was known to be made at this port, which good news, together with that of a general peace, will, I hope, make the Bengalees think of again reviving their trade to this place."

At first the only competitors with the British merchants seem to have been the Dutch, but in 1755 one "Monsieur Padree" arrives at Basra and is established as French Resident. The records

[1] Some of the finest arabs are bred on the Upper Tigris, particularly round Mosul. There is an important horse trade with Bombay.

of the subsequent years do not, I fear, hint at great cordiality between the representatives of the two nations in Mesopotamia, but as the factors were perpetually having rows with the Turks and each other, and there was hardly one whose career in that country did not close under a cloud of some sort, it is only fair to attribute this state of things in large part to the notoriously fervid climate.

The parent factory at Gombroon was in 1763 removed to Basra, which now became correspondingly more important; and in the following year the Ambassador at Constantinople announces that he has obtained from the Sublime Porte a *berat* for the creation of a Consulate at Basra, Mr Robert Garden being the first Consul. The Consulate at Bagdad was not established till 1802.

A terrible epidemic of plague swept Turkish Arabia in 1774, carrying off, it was estimated, two millions of the inhabitants. Before trade had begun to recover after this calamity another war broke out between Turkey and Persia, and the Persians once more captured Basra, and held it till 1779. The Wahabi Movement, meanwhile, from its cradle in Central Arabia, was rapidly spreading, and by the end of the century had reached the borders of Mesopotamia. In 1801 these fiercely fanatical "reformers" of Islam overran the country, and sacked the great Shia shrines at Kerbela, as they had already done with the shrines at Mecca and Medina. Peace was not restored till the Wahabis had been trodden under by the relentless heel of Ibrahim Pasha with his Egyptian troops.

As time went on the political, as distinct from the purely commercial, aspect of our connection with Mesopotamia became more pronounced, and in 1812 the post of "Political Resident in Turkish Arabia" was created by the Honourable the Court of Directors of the East India Company in conformity with a recommendation by the Government of India. Mr Rich first filled the post, residing at Bagdad, with Captain Taylor as his assistant at Basra. Before very long the inevitable conflict between economy and expansion arose. The Resident at Bagdad needed a large sepoy guard for his protection (once at least the Residency was besieged), and in 1834 a proposal was made to abolish the post and maintain only the Political Agency for the Persian Gulf at Bushire. Lord Clare, Governor of Bombay, pointed out the larger aspects of the question of our position in Turkish Arabia, and how deeply Indian interests were concerned. His view prevailed, and the Residency has remained till to-day. The official attitude towards it of recent years may be gauged by the fact that a large new residence for our representative was erected a few years ago at a cost of some £20,000.

The first efforts to establish a regular overland route for the Indian mail *viâ* Syria and the Persian

Gulf were made in 1834. For the purpose of this scheme Parliament voted a grant of £20,000 for the construction of two steamers to ply on the Euphrates. The necessary arrangements were made with the Turkish Government, and Colonel Chesney arrived at "Port William" (Meskeneh), on the Euphrates, and put together the boats which had been sent out in bits. Having completed this task, he made a survey of the river, and on 18th June 1836 addressed the following jubilant communication to the President of the Board of Trade :—

"I have the honour to inform you that this vessel (the *Euphrates*) reached the junction of the rivers during the afternoon of this day, so memorable for ever in the annals of England. We are now about forty-three miles from Basra, and have completed the survey and descent of the splendid river Euphrates. The officers and men are in good spirits, having arrived here without any difficulty or annoyance."

But the enterprise, alas! was not doomed to go through so happily, for during a heavy squall the sister ship — the *Tigris* — upset and foundered, drowning two army officers and twelve other Europeans. Three more steamers were soon after sent out, one of which, the *Assyria*, under Captain Felix Jones, made the first ascent of the Karûn river, surmounted the cataract at Ahwaz, with the help of rows of men on the shore pulling on ropes, and went as far as Shuster. It was not till 1888, however, that the Persian Government opened the Karûn to general navigation, when Messrs Lynch and Messrs Grey Mackenzie (still two of the most prominent shipping firms on the Shatt-el-Arab) at once took advantage of the opportunity and ran vessels up to Ahwaz. It may not be inappropriate to insert here a quotation from Lord Curzon's standard work on Persia, *à propos* of the great potential trade-route into the interior of Persia thus inaugurated, it being a question which particularly appealed to his attention.

"The great merit of the route when opened and organised, from a British point of view, will be that the cities and villages of West and South-West Persia—Dizful, Khurremabad, Burujird with 17,000 inhabitants, and with a surrounding plain of great productiveness, Sultanabad, the centre of the carpet industry, and their dependent districts, which are among the richest corn-growing lands in Persia—will be brought within easy access of the Gulf, whilst their inhabitants will thereby be drawn into the mesh of the Lancashire cotton-spinner and the Hindu artisan. Kermanshah with its 60,000 people, and Hamadan with 15,000, at present only served by the Turkish route from Bagdad, will also be brought within the southern zone of influence, and will swell the profits of Manchester and Bombay." Lord Curzon wrote this in the 'nineties, since when the preliminary survey for a railway line from Ahwaz to Khurremabad has brought the existence of such a trade-route a step nearer to realisation.

Since the dissolution of the East India Company at the close of the Indian Mutiny, British interests have steadily grown, both in the Persian Gulf and Turkish Arabia. The connection between our position there and the security of India has become more and more evident. We have long played the thankless part of policeman in the Gulf, to the great discomfiture of pirate, slave-trader, and gun-runner alike, and have followed a consistent policy of making friendly treaties with the chiefs of the Arab tribes which live along its shores. One of the more recent of these agreements has been with the semi-independent Sheikh of Koweit, a small town some sixty miles from the mouth of the Shatt-el-Arab, but of particular importance as being the one and only seaport at the head of the Gulf. With the Sheikh of Mohammerah, too, we have cultivated very cordial relations ever since he came into power.

Several new factors on a very large scale have been imported of quite recent years into the situation in Lower Mesopotamia, but so much has been written about them already that I will do hardly more than enumerate them. First of all, of course, the Bagdad railway. The concession for this stupendous scheme was signed in 1899, the year after the Kaiser's visit to Constantinople, but by the beginning of the present war the engineers were still held up before the great barrier of the Taurus, though the completion of the principal tunnel was reported in the autumn of 1915. At the time of our visit to Bagdad the line was being carried northward from there, and soon after reached Samara, seventy miles up the Tigris : the general estimate then seemed to be that at the rate of progress work was carried on at, the line would scarcely join up at Mosul in less than four or five years' time. The Bagdad railway possesses, of course, a whole literature of its own, and one that is by no means lacking in dramatic incident. The story of Germany's daring intrigues and plots to obtain a footing at Koweit—the ideal terminus on the Gulf—and their frustration on each occasion by the timely but unexpected arrival of a British gunboat in the harbour (as related in 'The Times' History of the War), has all the elements of a stirring schoolboy romance. The whole question of the railway, with its projected extension to Basra or the shores of the Gulf, is obviously one of those bewilderingly complicated problems with which our statesmen and diplomatists will be faced at the end of the war.

The oil question, although it was one of the principal factors in bringing about the Mesopotamian campaign, is in origin a Persian rather than a Turkish affair. In 1901 an Englishman obtained from the Persian Government a concession which gave him the monopoly of exploiting most of the oil-fields in the Empire. Eight years later the concession passed to a Company named the Anglo-Persian Oil Company, which started to work with great energy to develop its resources. Oil was "proved" in a number of different places, and the first important wells were sunk at Kasr-i-Sherin, rather more than a hundred miles east of Bagdad. The transport of oil from these wells, situated so far from any port, represented a serious difficulty, and their output has been only used till now to supply local needs; but when a supply was tapped at Ahwaz within reach of the Gulf, a pipe-line was laid down to the Shatt-el-Arab, and the great Abadan refinery, which I have already described, erected. The discussion in Parliament and the newspapers caused by the announcement that the Government, by the advice of the Admiralty, had acquired two million pounds' worth of shares in the company will probably be remembered by most readers. With the help of the new capital a fresh pipe was to have been laid down in order to double the output, when the war interfered. The old pipe was cut by the Turks early in 1915, but after the successful operations of General Gorringe's column from Ahwaz the enemy were driven far from the neighbourhood of the line and it was restored to working order.

A third event which, though it seemed comparatively fruitless at the time, may in the end have larger consequences than any, was the action of the Turkish Government some five years ago in calling in Sir William Willcocks to prepare a gigantic scheme of irrigation and drainage to include all Lower Mesopotamia. In spite of no small difficulties from provincial officials and unruly local Arabs, Sir William was able to complete the task. The estimated cost of carrying the project into execution was something like £20,000,000, a sum far beyond the range of the Turkish exchequer, and hardly likely to be contributed in the form of foreign loans considering the unsettled state of Irak. To convey some idea of the huge area involved, I quote from Sir William Willcocks's report on the irrigation of Mesopotamia. "The true delta of the Euphrates and Tigris comprises," he says, "the country traversed by the two rivers and their branches southwards of Hitt on the Euphrates and Samara on the Tigris." This means a distance as the crow flies of 400 miles from N.W. to S.E., and a superficies of some 12,000,000 acres, more than half of which, it is calculated, could be brought under cultivation. In view of the network of ancient canals which covers the face of the land, Sir William is careful to point out that "never in the history of Mesopotamia has the whole country been under irrigation at one and the same time." The volume of water, great as it is, would not suffice. Some of the old canals date from Babylonian times, others only from the prosperous era of the Caliphate; and nowadays, with the much deeper draught of vessels, it is necessary, of course, to keep a greater depth of water in the river-beds themselves. Some of the old works were, nevertheless, of astounding magnitude—witness the Nahrwan Canal, which dates from Chaldean days, and had a total length of about 300 miles along the eastern side of the Tigris. Its breadth, over long stretches, was from 100 to 150 yards, and its depth 40 to 50 feet—dimensions which exceed anything to be found in India and Egypt. This single canal carried such a quantity of water that it must, in Sir William's opinion, have seriously crippled the Tigris. The ruin of the Nahrwan was brought about by the Tigris changing its bed, owing to artificial influences, and striking eastward till it cut right into the canal itself, producing inevitably a fearful cataclysm when it did so.

The delta seems to have come to ruin and devastation less from neglect than from excess of zeal, the wholesale, unscientific irrigation of the soil having eventually dissipated the life-giving water and led to disastrous changes of level and the silting up of the proper channels. The waterlogged condition of parts of the Egyptian delta which necessitated such expensive drainage operations a year or two ago is proof enough how easily a virtue may be exaggerated into a vice even in modern systems of irrigation.

The scheme which Sir William Willcocks worked out for the regeneration of Lower Mesopotamia will, if it is ever fully applied, affect an area of over three million acres. To a great extent the old system of canals will be made use of, including the great Nahrwan. The scheme, of course, deals with very many problems besides the mere conveyance of water from one or other of the rivers to given tracts of desert. It includes the utilisation of huge natural reservoirs which, being connected with the rivers by "escapes," will store up the surplus volume of water which comes down in the spring spates when the snows have melted on the far-off mountains of Armenia, the construction of numberless "drains" to prevent waterlogging and the great waste of the marshes, and all the elaborate apparatus of barrages, weirs, regulators, dams, flood-banks, and so forth, that modern irrigation implies. Finally, it is proposed to keep up the level of the Tigris for the purposes of navigation—a most essential point, as it is already so low at certain seasons that only very shallow draught steamers can get up—by building regulating heads at the mouth of three branches which lead off it and are largely responsible for the trouble, their beds being lower than that of the Tigris itself. The opening of the Hindieh barrage in 1914 was the first-fruits of the scheme. The barrage diverts water from the Hindieh into the Hilleh branches of the Euphrates (the latter, which passes close to Babylon, had become silted up), and irrigates once more a large area of country which had long fallen out of cultivation. All the work connected with the barrage was carried out by a great English firm of contractors—namely, that of Sir John Jackson.

To come, lastly, to the modern successors of the factors of the East India Company, who were for so long the representatives of our nation in Mesopotamia. The Englishman of to-day whose lot is cast on the banks of the Tigris lives, one need hardly say, under far pleasanter conditions than those old pioneers who, according to the records, passed their existence in such a peppery atmosphere both in the literal and metaphorical sense. In each of the three towns which boast an English population, Mohammerah, Basra, and Bagdad, there is a flourishing English club (I am speaking, naturally, of normal times of peace), and Englishmen foregather to race, play cricket, tennis, and golf as they do all the world over. While at Mohammerah the Anglo-Persian Oil Company contributes the largest section of the colony, the Basra colony are chiefly merchants engaged in the shipping of dates. Half a million pounds' worth of this fruit in its dried state is exported on an average every year from Basra. The dates are not the dessert variety packed seductively in double rows in an oval wooden box,—those come from Tunis and Algiers,—but are mostly of the humbler sort which you see crushed out of all recognition on a coster's barrow. The trade is principally in the hands of Europeans, and the first shipments of the season from Basra are attended with almost as much rush and excitement as existed in the old days of the tea-clippers. At Bagdad the English are divided into a greater number of categories—bankers, merchants, shippers, missionaries, and a considerable number of engineers.

Although the average maximum temperature for some months in summer at Bagdad is something like 110° in the shade, and 120° is by no means unknown in some places, the country is decidedly a "white man's country." The only serious endemic disease is the "Bagdad boil," which attacks almost every inhabitant of that place sooner or later, and is particularly unwelcome from the fact that — being due to a mosquito bite—it very frequently occurs on the victim's face, and is apt to leave a scar for life; otherwise the country is healthy despite the tremendous heat, and it is remarkable how seldom you hear an English resident "grouse" about the climate. At Bagdad life in summer is made much more livable, thanks to the *serdabs.* A *serdab*, as the reader may already know, is a cellar ventilated by a shaft which runs up through the house itself and ends on the roof in a sort of masonry cowl so built as to catch the north breeze. When fitted up as a living-room, with comfortable furniture, it forms an ideal place wherein to æstivate.

When an Englishman wishes to travel from his native land to Bagdad, he has the choice of several routes. The ordinary, though prodigiously roundabout one, is *viâ* Bombay, the Persian Gulf, Basra, and so up the Tigris by river-boat. It takes about one month, and implies no serious hardships. It is possible, however, to follow a much more direct line overland by going from a port on the Syrian coast to Aleppo and thence driving to Bagdad, keeping more or less along the Euphrates; but the extraordinary discomfort of driving for many days on end across the desert prevents this way from being very popular. The caravan route from Damascus is hardly used at all by Europeans, as it means desert travel of the most arduous sort on a camel along an almost waterless road, and its only attraction lies in the interest of passing through the ruined city of Palmyra. There remain two more avenues of approach to the capital of Irak, one rarely, the other hardly ever, trodden by the *ferenghi:* the first is the river-route from Mosul, from which place the adventurous traveller sails down the Tigris on a raft of inflated skins, daring the risks of shipwreck or of puncture by an Arab bullet, the last the road from Kermanshah, which I am about to describe in the next chapter.

CHAPTER X.

ENTERING KURDISTAN.

OUR destination on leaving Bagdad was the little Persian town of Kasr-i-Sherin, to which point the Commission had, in the meantime, carried forward the frontier from Mendeli. Its distance from Bagdad is about 100 miles, of which the greater part —as far as Khanikin—is *carrossable;* beyond that point wheeled vehicles sometimes go, but it is not good to be in one of them. The Turkish mail leaves twice a week for Khanikin, and finding Captain Dyer, who had been making final arrangements for the Commission's transport, returning by it, I joined him as a passenger. The mail in question consists of a local variety of *diligence,* roofed, but open all round, drawn by a team of four mules harnessed abreast, and encased in tin—a feature which has earned for the type among the English people at Bagdad the name of "tin bus." Having filled the interior of our "bus" with kit, and bestowed ourselves at full length on top, we prepared to face the twenty-four hours' drive. The large *meidan* from which we started is the place where the caravans for Persia collect and load up, and a large portion of it was covered with bales and boxes awaiting transit. These were the accumulation of the past six weeks, during which time the road had been "stopped" — a very common state of affairs—by the turbulent tribes across the border. All the goods, of course, were degenerating, and the merchants' profits vanishing into thin air, but there was nothing to do but wait, as the only caravan which had tried to get through had been held up and plundered.

It was past midnight, and the moon had just set when we reached Bakuba and found, to our disgust, that, as the pontoon bridge is too decrepit for wheeled traffic, we had to turn out and ferry across the Diala in a *gufa*. After scrambling up the further bank, and walking for a mile in pitch darkness with the wretched porters staggering behind us under their loads, we found another vehicle waiting for us, a replica of the first. This time we had the postman and half a dozen mail-bags inside, so it was no longer possible to "spread" ourselves, and we had to share the narrow wooden seats with our new and odoriferous companion. Hour after hour we jolted and rocked and swayed across the interminable plain, over banks and through watercourses, with the old "bus" rattling and creaking like a ship in a hurricane. Presently the sun came up, and as its rays gained power, turned our tin-plated chariot into a veritable oven. It was quite the most purgatorial journey I ever experienced. The landscape was blank and empty, and besides the pilgrims the only living creatures to enliven the way were the bee-eaters. How I blessed the little fellows with their emerald breasts and coppery wings dropping in clouds from the telegraph wires as we passed, and bejewelling the ground as if to give us, at least, one touch of beauty! One never grows tired of watching them fluttering in the air, just like great big butterflies, then suddenly "planing" to the ground, with their graceful, slender bodies and fan-shaped tails as rigid as an aeroplane.

We stopped at the village of Kizilrobat for a change of beasts, but all the energy had been baked out of our limbs, and we rested supinely in the courtyard of the *khan*. Then we creaked and rumbled on again. As the day wore on, the mountains lost the unsubstantial look which the distance and the haze had lent them, and became a jagged line across the horizon. Then, towards evening, the endless plain broke into a ripple of low hills, on the other side of which we came into sight of Khanikin. Through the town flows a rapid river, the Elvend, which we crossed by an unusually fine stone bridge, the gift (if our driver's story was true) of a pious Persian dame whose charity took this practical way of helping the pilgrims. We found in the *kaimakam's* house our Turkish colleagues (who always manifested a quite natural preference for installing themselves on their own side of the frontier), and after dining with them, dragged our weary limbs to the *khan*. There, bedding ourselves on the roof to evade, as far as possible, the ravenous inmates of the place, we slept the sleep of the just.

Next morning horses were sent in for us, and we rode the odd twenty miles into Kasr-i-Sherin through a landscape of low hills, whose rugged sides formed a background for several grim, half-ruined castles, producing quite a north Italian effect.

Kasr-i-Sherin was our half-way house. Up to here we had been travelling through the plains, with the Tigris for our trusty ally ever within a few days' march; but from here on the mountains awaited us, and our communications would be both difficult and precarious till we reached the Russian frontier. All our spare kit had in consequence been sent on ahead in February from Mohammerah to Kasr, where we found it on arriving, and spent a busy time refitting and repacking. Big tents were exchanged for small, warm clothes stowed into the bottom of our *yakdans* against the cold of high altitudes, horses were shod, packs were lightened, and a hundred and one other preparations made for our new mode of life. In the matter of lightening our loads, the local inhabitants lent an enthusiastic hand. We had fondly supposed, on quitting the Beni Lam country, that the worst was over; but it soon became evident that the change was merely a leap out of the frying-pan into the fire. In spite of doubled guards and every other precaution against thieves, the people of Kasr stole our horses; they stole (poor man!) the Russian naturalist's collection; they even stole the bedding of the "Hajji Sahib." This last sounds like an anticlimax, but let me explain the circumstances of the case. The "Hajji Sahib," who was the chief of the Indian surveyors, was a gentleman of the portliest dimensions, who turned the scale at somewhere in the region of eighteen stone. One night he woke to the consciousness of something moving under him, and regained the full possession of his senses just in time to see the last of his blankets gliding out of the tent door. They had stolen his bed-clothes from under him!

Kasr-i-Sherin itself is a picturesque town clinging to the slope of a very steep hill, the summit of which is crowned rather theatrically by a large ruin showing patches of sky through its riven walls. The main street, which is also the bazaar, and is paved with boulders the size of cocoanuts, runs straight up the side of the hill, round whose base curls the swift Elvend, watering orchards of fig and pomegranate, fields of corn and lush meadows, with here and there a patch of purple-headed opium poppies. Flanked by this strip of verdure on either bank, the river winds like a vein of malachite through the landscape of reddish-brown hills, which stretch back to the big mountains beyond.

There was a big city here in Sassanian times, and the ruins of it cover a large expanse to the north of the modern town. Not much remains above ground, but there are two imposing structures which still defy the vandal picks of the local inhabitants who see in these monuments of their country's greatest era nothing more than stone pits whence to quarry blocks for their own wretched hovels. One of these relics is the aqueduct which brought to the town its water supply from a point on the Elvend some miles higher up, and the other is a large building said to be the throne-room of the Sassanian kings. The ground-plan is square, and each side is pierced by a noble archway—a symbol that the monarch was accessible for suppliants from all the four corners of his dominions. The walls are built of massive unhewn stones embedded in the hardest mortar (the inside of the arches being lined with bricks), and corniced off to provide a base for the great dome which must once have covered it. What scenes, I wonder, were enacted in that vast hall when Queen Sherin "the Sweet," the loveliest lady who ever shared the throne of Persia, visited the town which is still called by her name!

The frontier work at this stage was the most complicated the Commission had yet had to deal with. The line of the foothills, which had so long been our guide, had ceased to be so, and there was still a long distance ahead of us before the frontier attained the watershed of the main range, which would carry it with a few deviations all the way to Ararat. The intervening gap was composed of broken country lacking a clearly defined river system, and containing a good many scattered areas of cultivation, among which the frontier had to thread a cautious line. A considerable piece of territory had, too, been ceded by Persia to Turkey at this point, including the site of the Chia Surkh oil-fields exploited by the Anglo-Persian Oil Company, for which, of course, special provision had had to be made. Owing to the discrepancies of maps and similar causes, the Commission's duty of demarcating the line laid down at Constantinople was not always found to be feasible when we arrived at the spot. In one place, for instance, we were faced with the problem of making the frontier run along the crests of two hills which were discovered in real life to be *parallel to each other*—a poser which the most ingenious of geometricians could hardly have solved. At last, however, our immediate difficulties were settled, and we started.

Three miles from Kasr, where the road broke through the ancient aqueduct, eight loathsome vultures sat perched in a row on the crumbling parapet eyeing the carcass of a dead horse which lay stretched by the roadside below. We passed by this grim allegory of an empire's decay, and soon after quitting the main "road" to Kermanshah came to a place called Tang-i-Hamman, where there are hot sulphur springs. The hills all round were barren and treeless, and the fauna of the region, to judge by the specimens we saw, were as unlovely as their habitat. Once a slinking hyena disappeared over the hill-top at our approach, and later on we came on two fearsome reptiles, which might have been the young of ichthyosauri—scaly venomous-looking creatures about a foot and a half long. Later in the day a curious sight came into view. An endless wall, in ruins but clearly traceable, ran in a bee-line across the landscape. Where we crossed it it appeared as a tumbled mass of large boulders, through which a gap had been made for the path. We met it again several times in the course of the next few marches, and I afterwards learnt that the wall is said to stretch for a hundred miles or more. The Persians attribute it, like most monuments of antiquity, to the mythical Feridun, and I doubt if in reality anything of its history is known. One may, however, safely guess that it marked the boundary of some empire of the past which coincided remarkably with the new frontier we were laying down; and I was inclined to congratulate myself, on looking at the solid blocks, that I was not on *that* Frontier Commission.

Our camp that evening was pitched on a horseshoe ledge looking out across the Zohab plain. The long well-watered plain, mottled with cloud-shadows and hemmed in by hills on every side, with the mountains rising in tier above tier behind, was a beautiful sight. But the blight of anarchy lies on Zohab, as it does on so much of Persia, and the untilled soil produces nothing but wild oats and coarse tall grass except for a few poor patches of corn sown by the tribesmen.

Not many years ago there was a flourishing town of the same name as the plain lying near its eastern edge, but it is now completely in ruins. The surrounding country is the home of the Gurân, an important Kurdish tribe which has of late years rather come down in the world; and near Zohab, on a mountain called Dalahu, is the "holy place" of Zardeh, the shrine of the Ali Illâhis, to whom this tribe adhere. The Ali Illâhis are the followers of a very remarkable religion about which little is known. It contains the elements, at any rate, of fire-worship; but Soane[1] considers that "there is no guarantee that Zoroastrianism was the original faith, though there are strong traces of it": he thinks it is probably an agglomeration of the customs of a number of different religious systems containing a core of secret rites cloaked by certain orthodox observances for the sake of avoiding persecution.

Sir Henry Rawlinson, when he was Consul at Bagdad in the 'thirties, visited Zohab, and has left on record what he learnt concerning some of the principal tenets of the Ali Illâhis. They believe, he says, in manifestations of God in-

[1] Author of 'To Mesopotamia and Kurdistan in Disguise.'

carnate in the forms of 1001 persons of different ages, among the most important being Benjamin, Moses, Elias, David, Jesus Christ, and—as their name implies—Ali. Rawlinson identified them, apparently on rather slender evidence, with the Lost Tribes. Soane, on the other hand, quoting from Persian authors, suggests that they are of Mohammedan origin, having first come into existence in the lifetime of Ali. Ali himself took drastic measures with his would-be devotees, bundling them into a pit and having fire thrown in on top. Persecution, however, as so often happens, served but to fortify the persecuted, and the unfortunate heretics' only reply to Ali as the burning brands fell on their heads was, "Now it is a certainty that thou art God, for the prophet has said, 'None but God shall punish with fire.'"

Beyond the plain of Zohab lies the twin plain of Serkaleh connected with it by a narrow neck or *tang*, so that the two together have very much the shape of an hour-glass. The hills next day were canopied with clouds, and whisps of vapour floating across the plain filled it with a delightful play of light and shade. The grass, girth deep, was gay with cornflowers and clumps of wild hollyhocks—or some flower nearly akin thereto—with petals of a soft and crinkly texture like *crêpe-de-chine*, and of varying shades of red, mauve, and pink. The silky grass on the hillsides rippling in the early sunlight added to the beauty of the scene ; and everywhere nature seemed to be doing her best to compensate us for the dreary desert marches of the past few months. Passing through the *tang*, we emerged into the plain of Serkaleh, which proved to be more extensively cultivated than Zohab, so that we rode for some hours through ripening corn before reaching our camping-ground, beneath the shade of burly Bamu. "Burly" is really the only term to express the almost menacing boldness of outline of this fine block of mountain with its 1000 feet of sheer precipice running along the upper third of the whole of its eastern face. It is reputed to be a great hunting-ground for leopards, and our *shikaris* scoured it with rifles; but the only big game we came across was an inquisitive old bear, who came snuffling round near our camp one moonless night when it was too dark to go out after him. But though we *enjoyed* poor sport, we *provided*, alas! the finest *shikar* that the neighbourhood had known for years. Our tents had imprudently been pitched within a hundred yards of a Kurd camping-ground deserted not long before in favour of the summer pastures. But though the human occupants had departed, they had left behind hordes of their agile companions, who, scenting fresh prey, hopped across in their myriads to our camp and feasted royally on the exotic delicacies they found there. Life in that camp was Hades! No writer on Kurdistan, in my experience, has failed to remark on the number and voracity of its native fleas; and even Mr Soane, the ardent panegyrist of all things Kurdish, has to admit the charge, though he softens the impeachment against his *protéges* by maintaining the slightly ingenuous theory that the dust of the country breeds the pests spontaneously.

At no great distance from where we are camped there is a village called Pouchteh. An interesting rock-carving is visible near by, which was pronounced by the archæologist Sheil to be very possibly the most ancient known record of its sort in the whole of Asia. Accompanied by a villager as guide, we went to see it. After scrambling some way up the steep hillside without seeing any sign of the object of our quest, we began to fear that our guide was taking us on a wild-goose chase. Presently the man stopped and pointed upwards. "It is up there," he said. "But where?" we replied, for there was nothing visible except a blank wall of rock, nine or ten feet high, and almost perpendicular. "Climb up on to my shoulders," he said; so on to his shoulders we climbed, and by the help of some crevices in the rock managed to hoist ourselves up on to a narrow ledge running across the rock. We followed the ledge round the corner, and there came on the carving. It had been cut in the rock-face, a foot or two above the level of the ledge—which was barely 18 inches wide—and was thus completely hidden from any other point of vision than our own; in fact, any one not knowing the secret of its exact position might have hunted for it vainly for a lifetime. What considerations can have made the sculptor chose this concealed and almost inaccessible spot is hard to imagine, but no doubt its preservation (it is in excellent condition) is largely owing to his choice. The carving, which is in high relief, is perhaps a foot and a half square, and seems to represent a king armed with bow and arrow, and a prisoner of war in supplication on his knees before him, there being a cuneiform inscription in the margin. Unfortunately it is not possible to get far enough away to take a photograph. The local Kurds reported other and larger carvings in the neighbourhood, but we were not able to find them.

We had now penetrated well into the heart of Kurdistan, and the Persian and Arabic which had hitherto supplied our linguistic needs no longer availed us. None of us knew a word of Kurdish, and only three of our followers spoke the tongue, so we were more or less cut off from intercourse with the natives. Persian roots play such a large part in the language that we were able later on to understand occasionally the drift of a Kurd's remarks, but at first not even this was possible; and it was fortunate for our comfort that we were, as a party, so self-contained as to be almost independent of the people through whose country we passed. Their language, one is told, repays study, and certainly this would seem to be the case to judge by the expressiveness of some of their place-names. Opposite Bamu there is a very craggy mountain whose name is Shevaldyr, which was interpreted for us as meaning "Tearer of Pants," while a stony neighbour rejoices in the rather analogous title of "The Breaker of Nails."

As soon as the frontier had been brought safely to the top of Bamu, the Commission started on a three days' march to the Sirvan. The Sirvan is merely another name for the upper waters of the Diala, the river which we had already crossed twice on our way to and from Bagdad. It has a peculiarity common to many of the rivers we crossed. Old Father Tigris is an inveterate poacher, and drains a very large area on the Persian side of the frontier range, whose crest-line is not therefore, strictly speaking, the watershed. The rivers, such as the Sirvan, rising on the eastern slopes, have refused, one and all, to irrigate the land of their birth, but seem to have felt along the containing range of mountains till they found weak spots where, breaking their way through precipitous gorges, they flow westwards to swell the waters of the great river.

Our road to the Sirvan took us first over the shoulder of Bamu, and we marched for a day along its slopes, which were sparsely covered with oak-scrub, relieved here and there—where a stream came down from above—by an orchard of pomegranates in their full scarlet glory of blossom. Then came a day's march down the valley of the Zimkan, and on the third day a stiff climb over the shoulder of a mountain brought us into full view of the Shahr-i-zur plain. From our high point of vantage the whole great plain was visible, dotted with the many mounds which proclaim it to have been once crowded with the habitations of men in those dim far-off days, when a highly-civilised race reared its cities where now all is waste and desolation. Far below, the Sirvan coiled and flashed in the sun; to the north-west the plain spread out indefinitely towards the town of Suleymanieh; while to the north-east the snow-flecked crest of Avromân—now our immediate goal—stretched as far as the eye could reach. We zigzagged down the steep slope, so thickly grown with oak that only the music of the mules' bells floating upward apprised one of the caravan below, and reached the river's bank. Here the local Kurds who ferry passengers across, thinking we were in their power, repudiated an agreement made the day before at the rate of five shillings for each *kelek*, and demanded four pounds instead. To their great disgust, however, after a careful manipulation of the mule-loads, the caravan was able to ford the river, and though the water reached to our girth buckles, and the tents got rather moist, we crossed without mishap. We marched on next day to Halebja, a little Turkish town lying in a fold of the plain.

CHAPTER XI.

ALONG THE AVROMÂNS.

THE chief feature of Halebja is, or was till quite recently, Lady Adela. Adela Khanum, to give her her usual title, belongs by birth to the family of viziers of Ardelân, a Kurdish tribe which, though now considerably diminished, was a few centuries ago practically an independent sultanate, and still retains some of its old prestige. She married one of the chiefs of the Jaff tribe whose headquarters were Halebja. The Turkish Government, following their traditional custom of propitiating such of the borderland chiefs as were too powerful to be coerced by force, appointed her husband, Mahmud Pasha, *kaimakam* of Shahr-i-zur. This led to his being absent for a large part of the year, with the result that Lady Adela, who was a person of eminently capable and decided character, replaced him at home. She not only managed her own and her husband's private concerns, but also (if one may use the term) "ran" Halebja. She built a prison, law-courts—where she acted as president—and a noble bazaar, besides several fine houses, which redeem Halebja from being, what it would otherwise be, merely an overgrown, squalid Kurdish village.

At the time when this enterprising lady was in her prime, the whole district of Shahr-i-zur was completely in the power of the Jaff tribe. In coalition with a smaller and very warlike tribe called the Hamawand, they controlled all the roads, and without their permission it was next to impossible for a caravan to pass from Suleymanieh to Halebja without being plundered. Constantinople had no authority at all, and when a Government telegraph line was put up the tribesmen merely appropriated the posts and wire for their own private uses without a word being said. With the Constitution, however, a new order of things set in, and when we visited the town we found Turkish postal and telegraph services working with admirable regularity. Adela Khanum, moreover, had to our great disappointment passed more or less into retirement, her husband having died some years before.

She is a personality of such interest, however, and her status presents such a contrast to the ordinary conception of woman's position in Mohammedan countries, that I cannot forbear to quote from Mr Soane's book a description of his meeting with her in 1909. Mr Soane was travelling in Kurdistan disguised as a merchant of Shiraz, and he thus relates the event :—

"In the manner of Kurdistan it was a private interview, so I found no more than twelve servants, retainers, and armed men standing at the door. The room was long and narrow, two walls of which were pierced with eight double doors opening on to the verandah, the other walls being whitewashed and recessed, as is done in all Persian houses. The floor was carpeted with fine Sina rugs, and at the far end stood a huge brass bedstead piled high with feather quilts. Before and at the foot of this lay a long, silk-covered mattress, and upon it sat the lady Adela herself, smoking a cigarette. The first glance told her pure Kurdish origin. A narrow oval face, rather large mouth, small black and shining eyes, a narrow, slightly aquiline hooked nose, were the signs of it; and her thinness in perfect keeping with the habit of the Kurdish form which never grows fat. Unfortunately, she has the habit of powdering and painting, so that the blackened rims of her eyelids showed in unnatural contrast to the whitened forehead and rouged cheeks. Despite this fault, the firmness of every line of her face was not hidden, from the eyes that looked out, to the hard mouth and chin. Her head-dress was that of the Persian Kurds, a skull-cap smothered with rings of gold coins lying one over the other, and bound with silk handkerchiefs of Yezd and Kashan. On each side the forehead hung the typical fringe of straight hair from the temple to the cheek below the ear, and concealing it by a curtain of hair, the locks called 'agarija' in the tongue of Southern Kurdistan. The back hair, plaited, was concealed under the silk handkerchief that hangs from the head-dress. Every garment was silk, from the long open coat to the baggy trousers. Her feet were bare and dyed with henna, and upon ankle and wrist were heavy gold circlets of Persian make. Upon her hands she wore seventeen rings, heavily jewelled, and round her neck was a necklace of large pearls, alternating with the gold-fishes that are the indispensable ornament of the Persian Kurd and of many of the Persians themselves.

"A woman fanned her, while another held cigarettes ready, and a maid waited with sherbet and rose-water.

"As I entered Lady Adela smiled and motioned me to a seat beside her on the mattress, and gave me the old-fashioned Kurd greeting—

"'Wa khairhatin, wa ban i cho, ahwalakitan khassa shala' (You are welcome; your service is upon my eyes; your health is good, please God). . . . Rakish-looking handmaids, in flowing robes and turbans set askew, stood about or fetched scissors and tape for the silk cloth she was inspecting. A Jew of the bazar was displaying to her his wares, taking huge orders for all kinds of stuff, and squatted before her, taking notes in Hebrew on a dirty scrap of paper. The maids advised, criticised, and chose cloth and stuff for themselves, which Lady Adela would promptly refuse, or occasionally grant them, for she treated them remarkably well. The audience made remarks upon the proceedings, often enough chaffing

M

Lady Adela regarding her purchases, when she would retort in quick Kurdish with the best humour, every one joining in the laugh which not infrequently was against her."

Although Lady Adela's position was probably unique owing to a happy combination of rank and character, the freedom of her sex which it exemplifies is entirely characteristic of the social life of Kurdistan. The veil and all it implies is unknown, and the women are, for all practical purposes, as free as in England. My first walk through the streets of Halebja brought this vividly to view. In place of the black-draped ghosts which in other Mohammedan countries peep and giggle, or else bolt like frightened rabbits on the sudden appearance of a European, the good dames of Halebja sat and gossiped on their doorsteps just like Mrs Brown and Mrs Jones in any village at home, quite unperturbed by the passing of a stranger. The younger ones among them were strikingly handsome girls, a little Jewish in type, with a splendid bearing and an honest, frank expression, as different from the sallow, dark-eyed "beauties" of the harem as a healthy English country girl from a bedizened actress.

On leaving Halebja after a few days' stay we struck at last straight up into the mountains. The Avromân range forms at this point a mighty wall stretching without a break for the best part of 50 miles from the Sirvan gorge northwards.

Towards the Turkish side this wall is very precipitous, though buttressed by a number of low spurs separated from each other by deep-cut valleys, most of them well watered from the snows above. The villages lie all in the beds of these valleys, and a great difficulty was introduced into the Commission's work by the fact that the crest-line does not form the frontier.

Persian Kurds of the Avromâni tribe (a very ancient tribe claiming descent from the great Persian hero Rustam) have in years past crossed over the top and founded villages at the head of several of these valleys, which nourish Turkish villages a few hundred yards lower down; sometimes even a single village is half Persian, half Turkish. It was to deal with such complicated positions that the Commission had to climb up into these alpine regions, and as from this point what I have to tell deals only with the actual experiences and impressions of our journey, I propose to make use chiefly of my letters and diary.

Balka Jura, May 31.—We left Halebja this morning. For the first hour or two, while the caravan was still on the edge of the plain or among the lower hills, we passed through several Kurd villages. Each one was built by the side of a stream and surrounded by trees, chiefly pomegranates. There is usually an artificial pond, or rather basin, in the centre of the village under some specially large trees; the basin, which is fed with running water, is enclosed with a broad stone-coping about 3 feet high, with a ledge of convenient height on the inner side which serves as a bench for the village grey-beards, who sit there and smoke and gossip and watch their own reflections in the water—an enviable occupation in this heat.

The villages themselves were empty, and the people living in tabernacles of boughs just outside. I wondered if spring cleaning was going on, or if they were indulging in the "simple life." Neither was the case—they had simply run away from the fleas. A similar exodus takes place every year, I am told, during the "flea season," the wretched folk being literally hunted out of their homes by these outrageous parasites, whose numbers are only reduced to a sufferable level after the whole house has been turned inside out and every sort of carpet and covering spread for some days in the full blaze of the sun.

After five and a half hours of clambering up rocky ravines and traversing steep slopes we came into sight of our present camping-place. The last part of the approach gave one a delicious foretaste of the scenery we may expect in these mountains. The path drops quite suddenly over the edge of a narrow steep-sided valley full of fine walnut-trees; at the further end there is a noisy waterfall, and near by a splash of magnificent purple iris. The village itself is on the opposite slope, to which it seems to be growing like a fungus to a rock. The angle is so steep that the flat brown roofs project one above the other in tiers, the door of each man's house opening straight on to the roof of his neighbour down below, so that streets become a quite unnecessary luxury. A little tea-shop by the wayside, round which some local gentlemen in dresses of flowered cotton were grouped, together with the trees, the waterfall, and the rather pagoda-like effect of the superimposed roofs, gave to the whole scene quite an atmosphere of Japan.

June 1.—There was a decided nip in the air when we turned out of bed this morning. Our camp is nearly 5000 feet above sea-level, and the streaks of snow on the main range are not so very far above our heads. Although we have just entered June, may-blossom and wild briar are in full glory a little further up the slope.

The whole Commission went out this morning to set up pillars in the Tavila valley two miles from here. The sides of the valley have been terraced, the terraces planted with mulberry-trees and divided by lichen-covered walls much favoured by poppies, but the bottom of the valley is full of walnut-trees, some of them giants, with here and there a grassy patch as smooth as a bowling-green. The charm of the place, however, lies chiefly in the number of little rushing torrents as clear as glass which you meet everywhere. Numerous small canals take off from the main stream, and are led cunningly along the hillside to water the patches of cultivation which hang, as though by their eyebrows, 400 or 500 feet above the bed of the valley. Some of these canals finish by disappearing altogether over the ridge, whence they are carried down into some neighbouring waterless valley.

We reached Tavila at 8 o'clock, passing on the way the *tekkeh* of the Nakhshbendi dervishes. A most holy family of these dervishes lives here of such wide renown that pilgrims even from India come up to this mountain village to bask in the sacred atmosphere. They belong to an order which is by no means ascetic, and have a charming residence, an arcaded building with a row of poplars in front, and a fish-pond in their garden quite in keeping with their monkish profession. Like the Trappists, however, they keep the precept "Memento mori" perpetually before their eyes. I do not know if they dig their own graves, but the sepulchres of their predecessors occupy a most prominent position just by the front door. Some of them are marked by really beautiful carved headstones rising from among the clumps of purple iris which here, as everywhere in Kurdistan, grace the cemetery. Others of the departed dervishes rest beneath imposing cathedral-like structures made of hammered-out kerosene tins varying from 2 to 8 feet in height, whether in proportion to the age or rank of the deceased I cannot say. It is a striking example of how Oriental taste which can create objects exquisite in form and colour will at the same time tolerate the most extraordinary outrages on the artistic sense.

Having ridden through the roofed bazaar of the village, the Commissioners held a meeting at the *Karacol.* A peculiarity of Kurdish villages is that you never quite know if you are standing on *terra firma* or on somebody's roof. In the present instance there was nothing to warn us that we were on anything but solid ground until we came suddenly upon a gaping hole through which one looked straight down into a large room, with a fish-pond immediately below inhabited by enormous goldfish! Personally, I experienced much the same sort of shock as I imagine Korah, Dathan, and Abiram must have felt under very similar circumstances. By the time the whole Commission and its attendant crowd of Kurds was assembled on the roof it began to feel anything but safe, and I was not sorry to start off with a Turkish soldier and a local guide to explore the head of the valley. For some distance the path, which was about 4 feet wide, wound steeply up between houses, occasionally diving underneath some one's top storey. Poor "Archibald," led by my *sais*, Imam Din, was like a cat clambering up the tiles—there was no room to turn him round, so he had perforce to follow. After a while the gradient slackened, and I was able to ride him a couple of miles to the next village. There I came on a tea-house by the roadside where a circle of Kurds were sitting and sipping tea, with their rifles—most of them brass-bound muzzle-loaders — hung on a branch overhead. My guide and soldier evidently thought that the climb ahead demanded a preliminary cup, so we "dropped in." The scheme of a Kurdish tea-house is delightfully simple. Four stone walls form a square, round the inner side of which runs a ledge to sit on. An extra deep recess in one corner serves as a table, where a little wood fire burns to provide charcoal for the samovar. The owner's outfit consists of the samovar, two or three small china tea-pots, a large array of *istikâns* (which are glasses about 2 inches high pinched in at the waist and prettily coloured), an equal number of filigree spoons, and a metal basin for washing up. The clear stream running by outside provides every other need except the actual tea and the sugar, which is dispensed with a generous hand. You come upon these tea-houses dotted about in the most unlikely places, and their appeal on a warm day is wellnigh irresistible. We have a native Lipton just outside our camp, by the way, who has added a fresh feature by suspending a string of wares from the tree overhead, but he offers nothing more ambitious for sale than sugar loaves in blue paper covers and an empty champagne bottle which he has managed to get from the butler.

However, to return to the subject, after taking a "cup o' kindness," or rather two or three, with the rough but genial company, we continued up the valley on foot till we reached the neck, whence there was a fine view of the Sirvan gorge. I saw a village far below and had thoughts of going further, but my Kurd assured me that they were "bad men" on that side (which means that they belong to a tribe with which his is at feud), and would not be responsible for our reception, so we returned home bearing a big block of snow taken from a drift and tied up in the Turkish soldier's coat. He and the Kurd were both excellent fellows, and we conversed together on many subjects in a wonderful jumble of Turkish, Persian, and Kurdish.

My first impressions of the Kurd are fully borne out now that we see him at close quarters in his native mountains. He is certainly a fine creature. The men are picturesque-looking ruffians in their many-pleated trousers, embroidered "bolero" coats, coloured turbans, and armoury of weapons stuck into their belts. Here in the hills they carry, too, a peculiar short coat of enormously thick felt, fashioned with dummy sleeves and a hole beneath for the arm to come through — rather on the principle of an undergraduate's college gown. Although it is warm, sometimes even baking, in the valleys, the wind near the summit, coming off the snow-fields, is most intensely bitter, and the shepherds need these arctic garments as badly as their flocks need their fleecy coats.

The women are very handsome as long as they are young, and here more than in any Eastern country I am struck by the sudden jump from youth to old age. You see plenty of fine up-standing girls looking about seventeen or eighteen, and an equal number of withered and bent old hags, but it is almost literally true that there is no intermediate stage; I do not think I have met a single woman whom you would describe as "middle-aged." Their fashion in hairdressing may have something to do with this. They wear their hair hanging straight down and cut off square, level with the shoulders—a style as becoming to the black-eyed damsels as it is grotesque in the case of the wrinkled old dames. Another habit which needs a good deal of "carrying off" is their way of clipping a stud set with a turquoise through the right nostril of the nose.

By far the most noticeable thing about the Kurdish women, however, is their head-dress. It consists primarily of a cloth skull-cap with a fringe of blue beads round the edge, and this is plastered over with silver coins, the lady's wealth being thus easily discernible at a glance. You see pathetic little persons of five or six with a meagre string of threepenny-bits round their caps, and you see strapping girls of a marriageable age with heads completely covered as if with silver scales, not to mention an overflow of coins suspended on chains round their ears.

It is a relief to be among a people who do not treat their women like slaves, but give them practically the same freedom as the men. Here they seem to have little to do except stroll about in the woods, the younger ones doing nothing, the older ones spinning with a tangle of yellow wool round one wrist and a spindle in the other hand: it may be, though, that the Commission's presence has been made the occasion of a prolonged bank-holiday.

Biara, June 4.—This is another such valley as Tavila, with streams and canals everywhere watering woods of walnut and mulberry. Our camp is pitched in an orchard of the latter. This is less imprudent than it sounds, for the mulberries are not the red variety we have in England, which leave an indelible stain on everything they fall on, but a white sort. The former do exist, but are rare, and are called "king" mulberries, the white fruit being the commonest sort and grown in immense quantities in order to be sun-dried and exported. Though not luscious like the others, they have a very subtle flavour. The Biara valley, instead of winding about like Tavila, runs perfectly straight, and abuts right on to the main range, so that you see the whole height of it rising abruptly only three miles away—a fine sight.

We only took a couple of hours to come here, but the roads are beginning to get troublesome. We have not actually lost any mules yet, though Wilson (who was obliged to go off to Kermanshah to fetch money) lost two over a precipice, and the Persians have lost one or two. Cash, by the way, is one of the difficulties of travel in Persia—at least in the case of an expedition on such a large scale as our own. The only coin worth carrying is the kran, which, like the Indian rupee in the old days, is worth its own weight in silver, and so is enormously bulky, a 2-kran piece—value 8d.—being about the size of a florin. Several thousand of these make, as you can well suppose, a serious item of transport. On the road here I had my first practical lesson in mule-driving. Together with the *jemadar* I brought up the rear. A heavily and badly loaded mule kept us back and let the main body get far ahead, so that presently we found ourselves a little party of three, including one *charvadar* and two mules, one the beast with the heavy load of tent-poles, and the other with a sore back and no load at all. At every possible opportunity the loadless one clambered up the hill-side to graze. While the *charvadar* chased him his companion would also take to the steep and there deposit his load. Both of them had then to be caught, the unwieldy 9-foot tent-poles dragged back on to the track, the load repacked and hoisted up again on to the mule's back, which, taking place on a 2-foot track, with a couple of horses to hold and at least one of the performers a novice at the art, was a good deal less easy than it sounds!

Biara is arranged on the pagoda system even more markedly than Balkha or Tavila, and when seen from across the valley looks like an enormous flight of steps leading from the bed of the valley for several hundred feet up the side. At the bottom, just above the stream, there is an ambitious-looking mosque with a double tier of arcades and a species of lantern (in the architectural sense), roofed with the inevitable battered-out kerosene tins and ornamented with what, if I am not much mistaken, were originally manufactured as chair-legs. Having made the acquaintance of a very gentlemanly Kurd who could talk Persian, and was particularly attracted to me, as he explained, by the fact that I had a gold tooth (which he probably took for a piece of reckless personal adornment), I asked him to take me to see the mosque. There we got into great trouble with a venerable old *mullah* who was sitting inside, and much resented the entrance of an infidel. "Who are you?" he shouted angrily. "What do you want? What's your religion?" I elected to answer the last of his questions, and explained very courteously that my religion was a peculiar brand called *din-i-inglizi* (a rough translation of the "Church of England"). Of course he had no idea what I meant, but being a great religious authority did not care to admit his ignorance in the presence of the other Kurds who were now standing round; so with a few more mutterings and grumblings, he let us go through. There was nothing much to see, however, and when I came out I only wished that the Biara fleas had been as exacting in their discrimination against a Christian as had been the surly old *mullah*.

June 9.—The Commission has been delayed here for some time by difficult negotiations. The situation is so complicated at Khan-i-Guermela, a village a couple of miles further up the valley, that the resulting frontier-line runs up and down, backwards and forwards, along irrigation canals which you can easily step across. It seems a pity when there is an almost impassable mountain crest only four miles away—the finest natural frontier imaginable. However, people's rights have, of course, to be respected, and as the men at the head of the valley owe allegiance to Persia, there would have to be a cession of territory to alter things. From here on the frontier keeps to the crest-line. I have had two good climbs during our stay here. The first was a scramble up to the crest-line to get a view of the mountains on the Persian side, and Merivân lake lying in its fine valley. My guide was an enthusiastic Kurdish boy from the village, who, far from treating me as an innocent lunatic for wanting to climb to the top, as local guides usually do, urged me to other and more ambitious excursions, such as he said he often made for the mere pleasure of climbing. His only petition on returning was for an English razor for shaving his head. I noticed, however, that he carefully took me home by a rather roundabout route which led us through a village where many of his relations lived, and from the unanimity with which the whole population flocked on to its roofs to stare the moment we passed through, I judged that he had scored a success by warning them that he would lead a queer curiosity, in the form of a tame Englishman, through their village on that particular evening.

My second climb was a matter of business. So many pillars had to be erected in this section that the usual sub-commissions were exhausted, and the construction of one pillar—(the least accessible of all)—was entrusted to the corps of secretaries. The path was long and very steep, and my Turkish colleague proved himself a hardy mountaineer; but "the Beauty of the Empire," being inclined to stoutness, did, it must be confessed, droop a little by the way. We were met at the top by loud protests from an ancient Kurd from a hamlet near by, who repeated many times that the decision of the Commission was giving to Turkey part of the sacred soil of Persia, which had been hers for 1320 years (I failed to discover the basis of his chronology). Despite all his vociferations, however, the poor old fellow was eventually shooed back to his village, and with the help of a squad of Turkish soldiers a pillar of 9 × 9 feet was built, dimensions which (as its architects were careful to point out) constituted a record for the Turco-Persian frontier up to that point.

CHAPTER XII.

THE HEART OF KURDISTAN.

Gulambar, June 11.—The heat is intense. We are down on the Shahr-i-zur plain again, marching along the foot of the Avromân till we find a pass. This morning things were made worse by a prairie fire. It was started, I suspect, by some careless *charvadar* at the head of the caravan, and within a few minutes the short dry grass was crackling and blazing for a mile or more—the smoke from it has been visible all day. Evidently the peasants are used to fires of the sort, for I notice that all the stacks of hay are insulated by having a wide ring purposely burnt round them.

There are few places on earth prettier than Gulambar. It is a veritable jewel set in the midst of this parched ugly plain, and thoroughly deserves its name of "amber rose." Every rose has its thorn, though, and Gulambar is, according to all repute, the most snake-infested spot in all this country. Every one shuns it on that account, and our own camp is to be pitched some miles farther on. There have been several cases of scorpion bites in the camp already, so we are the more careful.

The dervishes are here too, and their head, Sheikh Ali, who resides at Gulambar, is a very big personage. His house is delightful. You enter into a courtyard containing a large stone-flagged tank, shallow, but all bubbling from the springs at the bottom. The rich brown stone of the house is reflected between patches of green water-plant, and the figures of half a dozen reverend old gentlemen in cloaks of the brightest hues sitting on a raised verandah behind completed a scheme of colour which was really exquisite. The houses of the village cluster round the minaret of a very old mosque, said to have been built in the days of Sultan Selîm, with a stork's nest on the top. On the farther side of the village, among groves of poplar, several copious springs burst from the rocks and supply a little lake edged with willow, and forming a perfect mirror for the old minaret and the stork family. The overflow—a beautiful clear stream full of trout and tortoises—runs through an avenue of poplars, in whose branches all the sparrows of Kurdistan seemed to have gathered. The ruins of a very massive bridge, such as the Romans built, still partially span the stream, and just below it there are hot sulphur springs walled in and marked by two or three upright poles covered with fluttering bits of votive rags. The sulphur smelt, of course, horrible, but it has had the curious effect of staining all the stones which line the basin into which it flows a most beautiful shade of blue.

Chakân Pass, June 12.—We took warning by our grilling of yesterday, and were on the move early this morning, getting in an hour's marching before the sun cleared the crest of the mountains. The track was a narrow lane between a wilderness of thistles and wild oats, with occasional patches of hollyhock. The Kurds harvest the thistles, which grow here very big and fine; but I do not know what use they make of them. Corn is grown in a casual sort of manner in the plain and threshed on the spot. We passed a good number of these open-air threshing-floors, which are nothing more than circular clearings where the corn is spread in a thick layer and oxen driven round and round to tread it out. Apparatus is reduced to a minimum, and the oxen, five or six in a team, are linked together with a single rope looped round their necks. They form a picturesque sight, and with the first yellow rays of the sun just grazing their backs and shimmering on the corn-tops the effect was Troyonesque.

During the morning we met Wilson, who had returned from Kermanshah by a different road, with his mules loaded with £2000 worth of krans—a fat haul for the brigands through whose country he passed if they had chosen to attack him. This is a particularly lawless region, as we are at the junction of four separate tribes, and there is consequently no one to enforce order. Last night robbers came into the camp, and our local guard blazed off a volley in all directions to show their zeal. Their headman hunted anxiously round the camp for blood this morning, but I don't fancy there were any casualties. We returned along the road by which W. had come, and after climbing several thousand feet (we had dropped down previously to 1900) camped near the head of the pass.

Pirân, June 16.—We have come into quite different scenery on this side of the Avromâns. Unlike the steep clefts of Tavila or Biara, the valley in which we are now encamped is of the comfortable full-bellied sort, covered with fields and oak-woods, and stretching up for miles in a straight line till it finally curls in towards the crest with a great sweeping curve which looks very like an old glacier-bed.

A catastrophe has occurred which has thrown a gloom over these pleasant surroundings. The doctor attached to the Persian Commission, a member of the last Persian *mejlis* and a most pleasant companion on this journey, went out partridge shooting yesterday. With his loaded gun in his hand he stumbled over a rock, fired the gun, and practically blew his hand off. He bandaged himself and returned to camp in the most plucky fashion, but Captain Pierpoint on being sent for had at once to amputate his hand. It is the right hand, and one feels that the loss is particularly sad in the case of a man of his profession, especially as he is one of the small number of European-trained Persian doctors. There is, of course, no possibility of sending him from here to any civilised place where he could get attention, so he will have to accompany the Commission in a litter till he is able to ride again.

Buava-Souta, June 17.—We have crossed yet another pass to reach this place, the Persian doctor standing the jolting of a well-meaning but very inexpert team of stretcher-bearers wonderfully well.

Not far from here there is a large storkery in the woods. Hitherto we have always found the storks nesting in villages, on the roofs and the walls and the minarets, but these seem of a less companionable nature, and have made themselves a regular town in the oak-trees. From a little distance off the colony has a remarkable appearance. The nests are simply great platforms, five or six feet in diameter, made of loosely woven sticks, and they look very like the tops of a battleship, while it requires a very small effort of the imagination to convert the motionless creatures standing in them into a look-out of bluejackets. On the ground underneath each nest you find a rather gruesome collection of dinner relics — dozens of empty tortoise-shells, and even bones. The old birds "lak-lak" defiantly at you till you get quite near, when they fly off, and occasionally an adventurous storklet peeps out at you over the edge. The same woods are full of beautiful blue jays and black-and-white woodpeckers, while the tortoises underfoot are a positive nuisance, even invading our camp at times. Brooke, who has a way of acquiring pets of all descriptions, has got from somewhere a brace of fox cubs, who live attached by long strings to a tent-pole. They are very tame and playful, and have the greatest respect and, at the same time, affection for Azaphela, the minute dachshund, who doubtless appears a veritable mastiff in their eyes.

It is perhaps a commonplace to refer to the human adaptability to new modes of life, but this journey is certainly an instance of it. We have now for nearly half a year been constantly on trek, moving camp, on an average, one day in three. At first one was consciously *travelling*, but I find myself reaching a stage now where to be on the move is the normal, and to stop for more than a day or two is the abnormal. I think one begins to get an inkling of how life appears to the real nomad. There is, however, a fundamental difference between us. Whereas the nomad more or less circles within prescribed limits, we continually go forward; and it is always to me a strange feeling to leave a valley where the Commission has camped for two or three days (and it is extraordinary how familiar it grows in that time), knowing that there are a million to one chances against ever seeing it again. The desire to revisit places is a curiously deep-rooted tendency, and I suspect that the primitive nomad has it strongly implanted in him.

Champeraw, June 23.—This little mountain village, with a name so suggestive of holidays in Switzerland, was the cause of much debate when the frontier was discussed at Constantinople. No one could say for certain whether it belonged to Turkey or Persia, and the upshot of it all was that it was left to the Commission in its wisdom to decide. The final judgment depends on what the position of affairs was in 1848 as revealed by our investigations on the spot. As both sides may be relied upon to produce octogenarians ready to swear that they passed their infancy under the benign rule of Shah or Sultan (as the case may be), one is tempted to make the nefarious suggestion that we should toss up!

Between Buava Souta and this place the frontier, following the crest of the Zagros, makes a colossal "hairpin" bend. After following round for about sixty miles, it comes back to within twenty of the point it started from. Fortunately there is a very sharply defined watershed, so our main party has been saved from having to "loop the loop"; and we have only made one day's journey into the bend for the sake of an *acte de présence* on the part of the Commissioners, and various sub-commissions have been told off to set up pillars on the passes. There is one point on the northern side of quite dramatic interest from the geographical point of view. It is the watershed between two small streams of equal volume but very different destiny. One of them arrives eventually in the Persian Gulf, the other in the Caspian Sea. A philosopher might well be tempted to moralise on the apparent insignificance of the great "dividing-lines" in nature.

The road which we followed down the great bend was very different from the narrow tracks we have grown used to in these mountains. It was a broad, beaten highway, fenced in on either side wherever it passed through patches of cultivation. It has been worn as it is by generations of the Jaff tribe passing through on their annual migration from the neighbourhood of Kizil Robat, on the Bagdad - Kermanshah road, to the high-lying pasture lands near Sakiz in Persia. They are the terror of the whole country as they pass through with their flocks, plundering and ravaging whenever they get the chance, and the local agriculturist makes a rather pathetic attempt to protect his crops by putting up a fence round his land.

On our return journey we followed along the banks of a river teeming with fish. The doctor, who is an ardent disciple of Isaak Walton, turned up in camp in the evening triumphantly carrying a 3½-pounder. Yesterday we fished all day with fair success: I caught a 2-pounder on a "spoon," and lost another of the same size in trying to land him at a difficult place. They are quite a sporting sort of fish, and are very fair eating—an important consideration when our commissariat is as low as it is at present. No one of our party can definitely identify their species; but I am told that, except for size, they resemble very much the great Indian *mahsir*. Our promiscuous collection of live stock was augmented in the course of our angling by a couple of young magpies. Their nest was in a willow by the riverside, and our arrival so disconcerted them that they fell out into the water, whence they were retrieved by "Ben," wet, scared, but unhurt. A suitable cage is being constructed for them under Brooke's auspices, and they will henceforth form an addition to our travelling menagerie of Persian fauna.

June 27.—We have moved our camp up to 6000 feet, and the view, as I write here on a little plateau in front of the tents at 6.30 in the morning, is exquisite. There is nothing but a great expanse of mountain-tops, with spaces of grey mistiness between. One can just catch the murmuring of the river 2000 feet below, but there is not another sound. Far to the north the white peaks of the Kandîl range are dimly visible, though still nearly three weeks' march away.

7.30 P.M.—As negotiations over Champeraw are still dragging, W. and I, wearying of the long wait, took this chance of a flying visit to Baneh. We failed to quite reach it to-night, and are camped by the roadside—that is to say, our camp-beds have been put up in a field near a small hamlet, whose inhabitants have come out *en bloc* and follow with the keenest interest every detail of our meal and toilet. The head-man came too; and W., who has the happy knack of getting natives into a good humour, has been raising choruses of laughter by a few simple jokes. "What's your income?" he asks the agha (this is quite as proper a question in the East as an inquiry after a stranger's health would be in England). "Ten tomâns," is the answer. "And your perquisites?" W. continues. "Oh, I never steal," says the agha. "Well, then, *you're* not much of a Kurd," retorts W.; whereat the crowd roar with laughter, for the Kurds quite openly plume themselves on their cleverness as thieves. Among the supplies which the villagers brought for our dinner was a jar of the sweetest wild honey. It is quite an important item in the diet of the people of this country; and in conjunction with crisp native bread (unleavened, and pancake-shaped like the

Indian *chupattie*) and a bowl of thick *mâst* (a species of junket), it makes a meal for a king. *A propos* of wild honey, W. tells me a story of a conversation with some nomad Lurs when he was living in their tents in Luristan. They were eating honey, and W. began to describe to his host how in England we keep "tame" bees in hives. But the 'cute Lur was not to be taken in with such a palpable traveller's yarn. "Ho, ho!" he said; "and I suppose when you go up to your summer pastures you drive them in front of you like sheep!" The laugh was on W., as the Americans say.

On our way here we stopped, just before sunset, to drink tea at a wayside booth. As we sat there a party of horsemen, escorting one of the chief men of Baneh, came riding by. The last glint of the sun, striking level over the mountain-tops, fell on them and added a finishing touch to the picturesqueness of the cavalcade. All the men wore immense turbans composed of three or four various coloured handkerchiefs intertwined with a fringe of tassels dependent all round the head, except one, who, being a Mukri, wore the distinctive head-dress of his tribe—a sugar-loaf cap, with a long tassel attached to the peak and tucked into a turban wound round the base. Their trousers were cut like a Dutchman's, very voluminous, and caught in tight round the ankle, and their cumberbunds of flowered cotton or silk so immense that the array of silver-hilted knives and pistols that each man carried was almost lost among the folds. For footgear they wore the regular country *givas*—a soft cloth shoe not unlike the *espadrilles* which the Basques wear, and the best thing in the world for climbing over rocks. There seemed to be an absolutely unlimited choice in the matter of colour and material for their coats. One had a jacket with alternate stripes of white and bronze, another a flaming chintz with red flowers and green leaves on a white ground, a third was resplendent in bright green satin; some coats were plain and some were quilted, and among the whole crowd of them there were no two even approximately alike. It was indeed a wondrous spectacle as they filed past on their betasselled steeds in the dying rays of the sun; but even under their outrageous "get up" one could not help noticing with admiration the fine features and bearing of these pure-bred mountaineers.

After they had passed, we wound on up the valley in the semi-twilight, reaching the top of the pass just in time to see the whole of the next range ahead flushed crimson by the sun which had set for us many minutes before. Slowly the colour died out as we started on the descent, and long before we reached our camping-place here the warm glow had given place to the beautiful but cold light of a crescent moon. Tomorrow we shall be under way before daybreak, and hope to reach Baneh in a couple of hours, whence a *kâsid* will carry the mails to Kermanshah.

Baneh, June 29.—We have had a disturbed night. Yesterday we announced our readiness to inspect any carpets which the good people of Baneh might have for sale. A dozen or so were brought from the village, and some attractive Ispahan rugs were laid out before our tent for more detailed attention next morning. W. imprudently had his camp-bed erected on the pile. It would seem that half the fleas of Ispahan must have travelled hither in the rugs, and during the night they changed their quarters for W.'s bed. He swears to having killed sixty! I myself escaped lightly, as my bed was inside the tent, and I fell prey only to a few of the enemy's scouts. One needs to see well to the tucking-in of one's mosquito-net in this country, for besides the comparatively innocent fleas, the place swarms with the most gruesome-looking tarantulas, and a very poisonous variety of centipede who leaves his claws inside you if you try to knock him off.

The town of Baneh, which counts about 700 houses and 7 or 8 mosques, lies in a broad, well-watered valley at an altitude of over 5000 feet. It is an example of a thoroughly Kurdish town, almost completely cut off from the rest of Persia, and paying not a sou, I believe, into the Imperial treasury. There was at the time of our visit a *karguzar*, or Foreign Office agent, a loquacious gentleman in regulation frock-coat, which he wore so as to display a large expanse of loud and very dirty shirt, dispensing altogether with a collar. His influence, however, was probably negligible, and the government resided entirely in the hands of the local *begzadehs* of whom the chief, one Mohammad Khan, was recognised as "Governor" by the Persian Government. Some ten years ago, being involved in a private feud, he invited in the Turks as his allies. A common historical phenomenon repeated itself, his guests remaining as masters and deposing their host. For six years the town was consequently Turkish, but early in the recent frontier negotiations it was agreed that the Turkish troops and officials should evacuate the place, and it reverted to the nominal sovereignty of Persia. The incident is a good example of the situation which has prevailed for generations along this frontier and has made our present work of delimitation so necessary.

Last night we interviewed Mohammad Khan. He was attended by a small company of notables and his mirza, or secretary, armed with a large pen-case which he ostentatiously laid in front of him on the carpet. The Khan's personal servant stood behind him periodically fitting a cigarette into a fine amber holder, quite 10 inches long, which he would then light and hand to his master. The *entourage* stood in a circle outside the tent staring hard at the unusual spectacle of two Englishmen. One had every reason to return the compliment, for some of their costumes were even more striking than those of the horsemen we encountered yesterday. I made a mental note of some of the colours affected by these gay gentry: mustard, heliotrope, crimson, and a really beautiful sort of old-gold were among the most striking. They wore the long sleeve which you see in various forms throughout almost the whole of Kurdistan. It falls from the elbow in a long point, so as almost to touch the ground when the arms are folded (like an exaggerated sort of surplice), but it is worn loose only in the presence of an important person, and at other times is twisted and tied tight round the wrist. There were several quite small boys in the crowd, looking particularly comical in their little tunics and enormous cummerbunds, with a couple of tags of white stuff attached to their shirts as make-believe sleeves. The cummerbund is, by the way, worn by some tribes in a most peculiar manner. It is twisted tightly into a sort of thick rope and then wound round the body, almost from the arm-pits to the waist; sometimes, instead of being wound straight round, it is criss-crossed in front somewhat in the fashion of a corset.

Moussik, July 1.—We rejoined the Commission here after a long march from Baneh. There are nothing but mountains between, and as our road cut across the grain, we never had a hundred yards of level anywhere. There is, however, water in abundance, and we were compensated by the most beautiful scenery. Trees of every variety seem to flourish; I noticed oak, willow, mulberry, wild-pear, sloe, and many sorts to which I could not put a name. We passed one black mulberry (or "Royal" as the Persians call it) apparently growing quite wild, but covered with the finest fruit. The lowest branches were unfortunately out of reach, and we could only get at the fruit by the expedient of my climbing on to W.'s shoulders and feeding him and myself alternately—an arrangement to which he took exception, on the ground that he had to trust too much to the integrity of his partner for his proper share of the ripe ones! We had our tiffin with us and ate it at a Kurdish tea-shop, sitting on stone seats round an artificial pool beneath the shade of a chestnut. The *ked-khuda* of the neighbouring village was sitting gravely smoking in the opposite corner, and accommodation was even provided for babies, in the form of a cradle. It was occupied when we arrived, but our appearance was apparently too terrifying, as the mother snatched up her infant and took to the hillside, whence we saw her creeping back to finish her cup of tea as we rode away. We are now in the tobacco country, and the lower slopes of the hills are planted pretty freely with tobacco-plant. It is cultivated on surprisingly scientific lines, the seedlings being reared in "nurseries" till they are a few inches high, and then carefully planted out between irrigation trenches dug with mathematical precision in a slightly crescent shape, each trench being 8 to 10 feet in length, and connected at either end with a cross trench running longitudinally down the slope. The state of the country is so unsettled that hardly any of the tobacco is exported, and I believe that the Persian Government get not a penny of revenue out of it; but seeing that tobacco is one of the most valuable of all crops, the potential wealth of this district must be considerable.

After many hours of riding through country which could only be adequately compared with a switch-back, we came late in the afternoon to a point not more than five miles from Moussik. Between us and the camp there was a big mountain, whose name is Sourkef, with the Zab running at its foot. Of course we supposed that there was a path rounding the mountain, so imagine our disgust when we found that the only way lay right over the very summit, a height of over 7000 feet above sea-level. It was nearly dark when we reached the top, but we could just detect the white tents far below us on the opposite side of a valley. There was neither water nor grass on the top, so we had no choice but to risk the descent. The path was a breakneck track barely a foot wide, the mules were nearly dead-beat, and a false step by one of them at almost any point would have sent him and his load to perdition. How the animals managed to scramble down by moonlight without an accident of any sort was a marvel, and I for one offered fervent praise to Allah when we got safely into camp at 10 P.M.

CHAPTER XIII.

KURDISH HISTORY, CUSTOMS, AND CHARACTER.

THE wanderings recorded in the last three chapters bring us to the very heart of Kurdistan. "Kurdistan" (like "Armenia") is one of those names which you find scored large across the map without any dotted lines or other marks to define their boundaries. The word simply means, in point of fact, the country of the Kurds, and as this people live in large numbers all the way from Adana on the Mediterranean to near Tabriz in Northern Persia (a matter of 600 miles east and west), the term is necessarily vague. In speaking of its "heart," therefore, I am not using the word in its anatomical, or rather geographical, sense, but as meaning the part of Kurdistan where the population is most purely Kurd, and which one may fairly look upon as the real stronghold of the race. Having conducted the reader thus far, then, it seems reasonable to give a few pages to facts of more general interest about the Kurds which may serve as a setting for our own particular journey.

The Kurds are, as a whole, a remarkably little-known race to the outside world. Their neighbours north, east, south, and west have all of them their own more or less recognised biographers. In Lord Curzon's 'Persia' we have an exhaustive description of the Shah's dominions and subjects; Lynch's 'Armenia,' on a lesser scale, fulfils the same purpose for the latter country; we are most of us familiar with the *bedouin* (those, at least, of Arabia proper, and they are not radically different from their cousins in Irak) from the pages of Sir Richard Burton, or one or another of his literary successors; while a whole library has been dedicated to the Ottoman Turk. Only the poor Kurd has been overlooked, or nearly so, and left to figure to the Englishman's imagination in the unique *rôle* of a bloodthirsty assassin weltering in the gore of massacred Armenians. That he frequently has weltered is, alas! undeniable, and the present war will certainly not help to cleanse his reputation in this respect, for the fate of the Armenians throughout his country has been one of its blackest pages. But we may still hesitate, I think, to utterly condemn this primitive race because of the crimes of religious fanaticism, more particularly when we remember that the record of our own civilised Europe in this respect bears such blemishes as the Eve of St Bartholomew. At all events in this chapter let us leave the "weltering" Kurd out of the picture and consider only some of his pleasanter aspects.

I must first, however, qualify the statement that no standard volume on Kurdistan has ever been written, by mentioning various books of travel in that country which do exist. During the course of the eighteenth century, Fraser, Millingen, Creagh, and Wagner all wrote accounts of their journeys among the Kurds, while quite recently Mr E. B. Soane has published a book describing his adventures there in disguise, which adds very materially to our knowledge of Kurdish customs, language, and literature. To these books I owe a considerable portion of my information.

There are various theories as to the origin of the Kurds, but there is one fact on which all the theorists agree—namely, the extreme antiquity of the race. The name by which we know it to-day goes back at least to Assyrian times, when the inhabitants of the mountains to the east of Assyria were known as "Kardu" by the Assyrians themselves. They seem to have been troublesome neighbours to the great Empire, and punitive expeditions were common, though apparently not usually a great success. Sennacherib, as Soane points us, is stated to have marched against a tribe in the Zagros Mountains called Kasshu and actually to have subdued it, particular care being taken to mention in the record that this was the first time that it had ever been conquered. There was evidently a very close connection, if not an identity, between these early dwellers in Kurdistan and the Medes. One authority states that they "coalesced" after the fall of Nineveh; another that it was only after the Medean Empire had passed its prime, and Ecbatana was no longer their capital, that the Medes withdrew into the mountains and founded the stock from which have sprung the Kurds of to-day. At whatever stage the Aryan invasion took place, there is very visible proof at the present day of the legacy which it has left among the mountain race, whose faces show not the smallest trace of Tartar or Mongol blood, as do many of the Turks, but bear, as I have already mentioned, a striking resemblance to the men of our own land.

I need hardly repeat the disagreeable experiences which befell Xenophon and the Ten Thousand when they passed through Kurdistan in the great retreat. The Carduchi, as they were now called, resented the passage of the Greeks through their country as they have that of all strangers since time immemorial, and expressed their feelings towards them by rolling gigantic boulders down on to the top of the wayworn army.

Coming to medieval times, when the Kurds were, beyond question, substantially the same people as they are to-day, we meet with the hero of the race—Saladin. That worthy antagonist of Richard Cœur de Lion, and greatest of the Sultans of Egypt, came of the Hakkâri tribe who inhabit the country to the south of Lake Van. As was natural, the Kurdish chieftains profited by the rise of their kinsmen to a position of such power, and many of them were established as rulers from Syria to Khorasân. The period was, in fact, the brightest in the whole history of Kurdistan, with the exception possibly of the reign of Selîm I. of Turkey (the conqueror of that very sultanate of Egypt which Saladin had raised to such a height), whose Vizier, the wise Idrîs, was also of Kurdish origin, and, like his predecessor in the land of Pharaoh, remembered his brethren in the days of his prosperity.

The modern history of Kurdistan is the history of its separate tribes, and it is therefore impossible to give any account of it in a short space. Several of the tribes have, however, achieved particular prominence, of which I may single out two, the Hakkâri and the Ardelân. The Hakkâri, who, as we have seen, boast Saladin as one of their tribe, flourished again in the sixteenth, seventeenth, and eighteenth centuries under their chiefs who bore the title of Princes of Bitlis. This line of rulers came to an end by the capture of the last chief by the Turks early in the nineteenth century. The chiefs of the second of these tribes, the Ardelân, held for some centuries a practically independent position as Wâlis of Senna, ruling over a small kingdom which included among others the provinces of Avromân, Merivân, and Baneh. At Senna they had established a little court where the gentler arts, such as poetry, were cultivated. In the middle of last century one of them married the daughter of Fath Âli Shah, a lady of as vigorous a character as the famous Adela Khanum of Halebja. This princess herself ruled at Senna for some years, but on the death of her son the Persian Government managed to break the succession and installed a Governor at Senna, which has now come largely under the authority of Teheran.

The language which the Kurds speak, "Kurmanj," as it is called, is usually described as an old Persian patois intermingled with many words of strange origin, ancient Chaldean for instance. Mr Soane, however, who has made a profound study of it, pronounces it to be a complete tongue of the greatest antiquity, possessing a rich grammar, especially in the case of the Mukri dialect. There seems even to be a small literature, which includes the original work of several Kurdish poets who wrote chiefly under the patronage of the little court at Senna. Old folk-songs one would, of course, expect to find among such a people, and I venture to transcribe from Mr Soane's book a translation into English of a very characteristic example :—

> "I would across the hills and far away, wife—
> Say shall I go or shall I stay, wife ?"

> "If you would go, God guard you on the track,
> And I will watch you from the pass till you look back ;
> I shall stand there in the sun until your clothes are shining white,
> Till you overtake the pilgrims that are travelling towards the night.
> What like of wife am I, if I weep or wail for you,
> Or leave neglected home and field to make a child's ado ?
> Christian, Turk, and Persian whimper thus and fear.
> Come, kiss me, and go swiftly, man and Mukri—ah, my dear !"

The character of the Kurd has been very aptly compared by Fraser to that of his own people in the highlands of Scotland. He says of them: "They are as devotedly attached to their mountains as any Scotch or Swiss highlanders for their lives can be. Like the first, they are divided into clans, acknowledging the supremacy of chiefs, who are regarded with as much devotion and followed with the same blind zeal, and all on the same grounds, protection, and kindness in return for fealty and service. They are proud, haughty, and overbearing, exactly in proportion to their ignorance, and like our highlanders of old, despise more or less all arts but those of murder and plunder—all professions but those of arms. They have their feuds with their neighbours, and make raids on the poor industrious lowlander; and woe betide him who should attempt to 'ascend the pass of Ballybrugh' or enter the 'country' of any Kurdish 'Donald Bean Lean' without guide and safe-conduct. The same love of enterprise and plunder has been inherent in both; each nation has been stained by like atrocities and fearful instances of revenge, and has been famed for the profession, and generally for the practice, of hospitality—the same regard for word once pledged—for the sanctity of the promise of protection."

To sum up, the Kurd is a loyal clansman, a pretty fair rogue, and a born fighter. The last characteristic is perhaps the one which strikes a stranger most. Dr Ross, who was the surgeon attached to the Bagdad Residency early in the eighteenth century, and visited the then all-powerful Mir of Rewandûz to attempt a cure for his father's blindness, says of them: "The element of the Kurd is war. He is trained to it from his cradle, and is never happy but in skirmishes and battles; I have seen boys of twelve and fifteen suffering from the most severe wounds received in recent fights. I understand their battles are very sanguinary. They begin with their rifles, but soon come to use the dagger all in earnest." The modern Kurd, hung about with a whole armoury of weapons, is a sufficiently strange and picturesque sight, but there are few things which I regretted more when travelling through his country than the disappearance of the lance as his chief arm. It was a long lance made of bamboo, having a large tuft or ball of wool concealing the point, with the object, it is said, of frightening the opponent's horse at close quarters. As a defence against the enemy's lance they carried, too, a small round shield of leather, and a certain number of them at least wore armour, either chain or plate, the latter in four parts made to fit the breast, back, and two sides. Fifty years ago it must have been a wonderful sight indeed to see a Kurdish tribe following their chief through the rocky defiles of their mountains on their way to raid the "lowlands" or to wipe out a blood-feud with a neighbouring clan. Fraser has given us a very graphic description of the spearmen of his day:—

"Both yesterday and to-day I rode out in the forenoon with the Khan's son through the Ooshoo (Ushnu) valley to have an opportunity of observing something more of my Kurdish friends and inquiring into their manners and customs. It was an animated sight to see the young fellows who accompanied us careering along the plain at full speed, as free, it seemed, as the beasts they chased, winding and turning their well-trained horses as they went through their spear-exercise to amuse me. Truly yon mounted Kurd, as he flies along as steady as a rock upon the spirited little animal he bestrides, is a gallant object; his splendid turban gleaming in the sun, his wide mantel floating away behind him, and the long slender spear quivering like a reed as he shakes his bare arm, the loose sleeves of his tunic and jacket waving like streamers in the wind. I assure you that as a dozen of these wildly picturesque figures were careering around us, I felt myself and my three or four kizilbashes in our sober garbs cut but a sorry figure. On they would come up to your very breast, their keen steel spear-points glittering like sparks of lightning, when a swerve of their arm or their body, imperceptible to you, would send them just clear of you, to wheel round with the speed of thought upon the other flank. It was a delightful representation of the admirably described combat between the Scottish knight and the Saracen warrior in the 'Diamond of the Desert,' which opens that delightful tale 'The Talisman,' and strange enough it seemed to find oneself amidst the very people there described."

Such scenes are, alas! a thing of the past, for the modern ·303 lends itself less readily to harmless display. The tradition has lingered on, however—or so I take it—in a peculiar Kurdish game which is common throughout the country, and which, by a rather curious coincidence, was played for our amusement by the son of the Khan of Ushnu, a great-grandson probably of the companion to whom Fraser refers. The game is of a very simple and primitive nature, and sounds somewhat insipid when described, though to the onlooker there is a decided satisfaction in watching the skill of the performers. As smooth a piece of ground as can be found is chosen, and the horsemen, riding singly at full gallop, when they reach the spot, throw a stout stick, some three feet long, point downwards, on to the ground a little ahead of the horse. The stick, if thrown properly, rebounds into the air, and the horsemen's skill consists in making his stick bounce up in such a way that he can either ride right underneath it or catch it in the right or the left hand. Needless to say, the trick is far more difficult when the stick is thrown on the off-side.

In matters of religion the bulk of the Kurds are, according to most travellers, bigoted *Sunnis*, very much under the influence of their sheikhs. There are, however, certain *Shia* tribes on the Persian side, and also a limited number of non-Moslems, including the Ali Illâhis in the south, whose tenets I have already mentioned in a previous chapter, and the Yezidis. The Yezidis are of a peculiar race of their own, and it seems a little uncertain whether or not they can be rightly classed as Kurds; their habitat is to the east of Mosul, and their beliefs and rites, which are very obscure, have been the subject of a certain amount of controversy. They are often, though some writers assert wrongly, designated devil-worshippers, and a sacred imitation peacock, which passes from hand to hand among the priests, held an important place in their cult, till the Vâli of Bagdad a few years ago led a campaign against them, secured the sacred bird, and sent it to Constantinople. I believe, however, that it was eventually restored to its worshippers.

Veneration of sacred places is a particularly marked feature among the Kurds. In the barest districts, where the woodman's axe has wantonly denuded the entire countryside, you will often come upon a single tree, or, it may be, a clump of trees, evidently of great age, spared on account of some pious association. Sometimes it is just a "pillar of stones" that you find set up to mark a hallowed spot—a custom at least as old as the days of the patriarch Jacob. The *imamzadehs*, or domed tombs, on the other hand, which are so common a feature further south and throughout Persia, you rarely see in Kurdistan.

There is a good-sized wood near Baneh, covering the whole side of a hill, and providing a home for a numerous colony of storks. As the whole of the surrounding country has been almost completely cleared of timber, one is naturally led to inquire how it is that this particular wood has been preserved. The reply you get is that the wood contains the tooth of Suleyman Beg. Suleyman Beg, it seems, was a famous saint who lived many years ago, and the good folk of Baneh, having obtained possession of his sacred molar, buried it with pomp and ceremony on the hillside opposite their city. The ground in the neighbourhood thereupon became inviolable, and to this day no man has dared to cut down a tree for a mile around.

Cemeteries one may almost call the *specialité* of Kurdistan. Would that we had some happier sounding name than "cemetery" to call them by —its dreary associations are so entirely foreign to the pleasant resting-places of the Kurdish dead. The truth is that the Kurds, instead of aiming at tucking away their ancestors as far as possible out of sight, always choose the prettiest spots in the landscape to lay their ashes. Often the graves occupy one of the isolated tree-clumps which I spoke of a moment ago, the grey headstones sticking up at all odd angles beneath the boughs like some strange sort of undergrowth.

The stones themselves are always carved—not with a dull recital of the dead man's name and attributes, as in an English churchyard, nor even, as a rule, with a scroll of koranic verses such as you see on Turkish headstones, but with a variety of queer formal designs whose significance, if they have any, is hidden from the passing stranger. The commonest design, so far as my experience went, was what looked like a conventional sun sending out rays on every side, or else a shell-like device of spirals and wavy lines. It would be interesting to know what part survivals of Zoroastrianism play in these old carvings. In some parts of Kurdistan there are conventional signs carved on the tombstones to indicate sex, rank, &c.,—a two-sided comb, for instance, for a woman, a dagger for a man; while above the graves of men of holy descent an iron hand is affixed,—a symbol, I believe, of the handing down of Imam's flags from one generation to the next.

It was in the mountains of Avromân that I was particularly struck by the beauty of the burial-grounds. They were usually under the shadow of fine oak-trees, and planted thickly with iris which at the time of our passing were in full bloom. A particular grave was often enclosed by a low stone wall along the top of which a row of pathetic little ornaments had been arranged by loving hands—chips of coloured marble, round pebbles or flags improvised out of sticks and shreds of white cloth. Ibex horns, too, are a very favourite adornment for graves, and you usually see a pair of them surmounting the sepulchre of a tribal chief.

Although I never actually witnessed a funeral in Kurdistan, I came one day on a scene of woe which left a very vivid impression on my mind. I was riding alone along a very remote path in the mountains when I came suddenly on a house built on a terrace on the hillside. A group of mourners, evidently returning from a burial, were winding up the steep. They were all old, grey-headed women, arrayed—like French soldiers—in red trousers and blue cloaks, and as they came they uttered the strange hysterical wail which is the mourner's cry throughout the East. Standing all alone in front of the house was a young woman, obviously the widow. She had strips of rags hanging from her arms and hands which she held stiffly out of each side of her. One by one the old women arrived at the crest of the slope where the widow stood waiting, and as each one reached her she fell on her neck and wept. For a minute or so the two women, the young and the old, remained clasped to each other, head on shoulder, rocking to and fro and mourning very loudly and bitterly. Then the elder woman passed on and the same scene was re-enacted with the next. I felt a horrible intruder on their grief, but none of them spared even a momentary glance at the foreign stranger riding by.

Fraser, who, owing perhaps to his highland origin, took a particular interest in the existence of superstitions, beliefs about ghosts and so forth, among the Kurds, and was very pertinacious in his questionings of them on the subject, declares that they are singularly lacking in any sense of the supernatural, and give little credence to *jinn* and suchlike beings. Without wishing for one moment to pit my slight experience against that of so careful an observer as Fraser, I must relate an incident which happened to me near the Zâb, and is evidence rather in the contrary direction.

We had arrived at our camping-place not far from a small hamlet, and I was wandering around while the tents were being pitched, when I came to a small walled pond full of clear spring water with a dozen or so of fish swimming about in it. Some of them were fine big fellows of a pound or more, and it was clear that they had not got where they were by natural means, so I asked some Kurds who were standing around whose they were. The answer, as delivered by my Persian servant who knew a little Kurdish, was to the effect that they belonged to an Imam, but he was dead. This was good enough, I thought; so having some fishing-tackle in my pocket, I baited it with lumps of bread and soon had three fine fish safely on the bank. The Kurds, of whom there were now a fair number gathered around, looked very askance at this, and kept saying reprovingly that the fish were "shakhs," which I mistook for "shakhsy" (meaning personal or private), and replied through my interpreter that if only the owner would show himself I would pay for the fish. This was evidently off the point, and it was only when another and more proficient interpreter appeared on the scene that I was brought to realise the enormity of my crime. My victims were not *personal property* but *persons*. In fact, they were the Imam himself. The holy man had died—I don't know how many years or centuries before—and been buried near this spot, and his soul, by some unexplained feat of metempsychosis, had passed into the fish. By the time I had grasped all this the poor creatures were dead, and it was too late to repair the crime; so, as it seemed a pity to waste them, they were cooked, souls or no souls, and eaten—and very good they were. The Kurds, I think, were divided between consternation at our appalling act of sacrilege and a sort of half-guilty amusement of the audacity of it. I suspect, too, that they were very curious to see what would happen to people who dared to eat an Imam. If only I had possessed some knowledge of Kurdish the incident might have been productive of interesting revelations of the Kurd's metaphysical beliefs. As it was, I failed even to discover how the fish got into the pond, or how they managed to exist in such narrow quarters once they got there. No immediate judgment of heaven, I may add, fell to avenge our guilt.

CHAPTER XIV.

FROM THE ZÂB TO USHNU.

Near Kandôl, July 5.—We look from here straight down on to the Zâb at the point where it breaks through the range. A little downstream from our last camp there were the ruins of a bridge—that is to say, the piers were standing but there was nothing to connect them with each other. The Zâb being a snow-fed river, liable to sudden spates, it is quite likely that the original bridge-builders deliberately omitted the arches or anything that would tend to dam the stream, leaving it to the local Kurds to throw across a temporary wooden structure which could be easily replaced. If so, no one seems ever to take the trouble to do so, and we found the bridge as little use in getting us across as Fraser had seventy-five years before. The caravan had consequently to ford the river, and though the spring rise had abated just sufficiently to make this possible, the rush of the water was still so great as to make it a stiff struggle for the beasts.

Our camp is now on a tree-covered ledge on the

hillside, 1000 feet above the bottom of the valley. We had meant to stop at Kandôl below, but two of us coming on ahead of the rest with the *pishbar*, and catching sight of this delectable spot far above, led the way up. When the rest of the party arrived, very hot and dusty, and ready for tiffin at the expected halting-place, and saw the familiar white specks perched high above their heads, their observations were, I believe, hardly printable; but we are all thankful enough to be up here to-day, with the thermometer at 106° inside the tents, and a first-class dust-storm raging in the valley below where our poor colleagues are sweltering. There is a waterfall, or rather nearly perpendicular water-chute, 100 feet high, near the camp, which adds an amenity to the situation, and provides a splendid natural shower-bath, and we get quantities of the most luscious sort of black mulberry. Indeed, we should have little to complain about but for a horde of tarantula spiders—great creatures 4 inches across, armed with a double set of the most fearsome mandibles — which have invaded our camp. They take a particular delight in crawling up the side of one's tent, and the customary rubber of bridge last night was entirely ruined by the sudden appearance of one of these monsters over the edge of the card-table.

Serdasht, July 10.—Serdasht is a small Persian town, 15 or 20 miles from the frontier. It lies off our line of march, so W. and I have ridden over to visit it and see something of the intervening country. It must have been a town of some importance once, but is now an insignificant place, largely in ruins. Compared with Baneh, it is in far greater subjection to the Persian Government, having a detachment of 100 *sarbâz* from Tabriz to act as garrison. We are here on the border-line between the two Persian provinces of Azerbaijan and Kurdistan (the latter word having also this restricted sense as an administrative area in Persia), and from now onwards we may expect to find far more evidence of Persian control, at any rate in the lower-lying districts.

Last night, on our way here, we slept under the stars at an altitude of 7000 feet, and woke up in the morning to find it nearly freezing. Our halting-place was a high alp covered with luxuriant pasture, and we had as our neighbours an encampment of Kurdish shepherds, besides a flock of very fine sheep, some buffaloes, and, to our surprise, a flock of geese. I only hope that the mess-secretary will have his eyes open when the main party comes through to-morrow! The Kurds had been anything but friendly on our first arrival, and warned us off their alp with very little ceremony. We were a party of only seven or eight, so had to use diplomacy, particularly as our dinner depended on the complacency of our hosts. After considerable trouble we got their *agha*, a most villainous, one-eyed ruffian, to come out and parley. We then discovered the cause of their ill-will. A quite insignificant quarrel about some fodder had taken place the day before between a couple of the Russian cossacks and some local Kurds, and the report of it had reached this out-of-the-way nomad encampment in such a distorted form that the men believed two of their comrades to have been killed. We had been taken for Russians (who are the only Europeans that most of the Kurds have ever heard of), and might have fared very badly indeed if we had not been able to establish our nationality, for no amount of talking would convince the Kurds of their mistake about our Russian colleagues. In fact, it needed all W.'s persuasive powers to banish the scowl from our *agha's* face, and it was only after several tots from our bottle of *arak* (the most serviceable weapon you can carry in Kurdistan) to finally dissolve it into the genial smile which told that the situation had been saved. In the meantime W.'s Indian orderly, a born diplomatist and a most invaluable asset on these sort of occasions, had got his fellow-Mussulmans into a thoroughly good humour, and assured our dinner, which we ate in the centre of a ring of the *agha's* men, now completely reconciled, and full of the usual talk about local "politics."

These nomads of the frontier are largely fugitives from justice, and all of them pretty tough characters. They are fine, big fellows, armed to the teeth, rifles in their hands, automatic pistols and daggers in their belts, and anything up to three bandoliers, crammed with cartridges, slung round their bodies. Any one of them will give his last farthing to procure a modern rifle, and you see every sort of weapon, from smooth-bore muzzle-loaders five feet long, with an iron crook to plant in the ground and act as a support for the barrel, to the latest pattern Mausers, the price of which in this country ranges up to anything short of £20. Their pipes, which they carry along with the arsenal in their belts, are of a curious type. They consist of a hollow stick 18 inches long, with a tiny metal cone for the tobacco at one end and an amber mouthpiece at the other, the size and shape of a bantam's egg. An *agha*, or other important personage, will never light his own pipe, but leaves it to his servant to fill and get going in full blast before he himself will deign to puff at it.

Vezneh Valley, July 18.—The Commission has suffered a dire blow. Mr Wratislaw has had to return home on account of his health, and he left this morning for Tabrîz, accompanied by the doctor. All four Commissions united to escort him for a mile or two on the road—an imposing cavalcade, for there is only a narrow path with swamp on each side down the valley, so that we had to ride most of the way in single file. They parted from us on the top of the ridge of hills which hems in the valley, and struck off north and homewards while we returned to our tents. I do not think there is a man in camp, from the butler to the *bheesti*, who does not deeply mourn the loss of the "*Burra Sahib.*" The duties of Commissioner will fall from now onwards on Captain Wilson, and those of Deputy Commissioner on Colonel Ryder.

Khanieh, July 25.—We have come here in two long marches from Vezneh. I have never, I think, seen more beautiful scenery than the country we passed through yesterday. It was a cool morning, with a fine wholesome breeze blowing, and the feel of an English autumn day. Fleecy white clouds were cruising merrily in the sky, trailing shadow-patches across the landscape, and the distant views were soft harmonies of brown and grey, such as you see in the Lake district. We were riding through the last of the wooded part of Kurdistan before entering on the treeless waste which stretches from here to Ararat. As though doing their best to make up for this, the trees were magnificent, and much of the road lay through splendid forests. After topping the pass we began to descend towards the Zâb. Ever since we crossed it at Kandôl eighteen days ago we have been marching parallel to it, but I have only seen it once, from a distance, when W. and I went to Serdasht. The far-off slopes across the river, all chequered in yellow and brown with the ripe corn-fields, and toned to the softest colours by a light mist, seemed like bits of another world when you caught a glimpse of them through a gap in the tree-tops. Here and there between the woods there were hay-meadows, some of the hay already mowed and stacked in heaps, and the rest shimmering like shot silk in the early sunshine, and jewelled with wild-flowers. There was a kind of long-stalked pimpernel, masses of cornflowers, and every now and then a clump of the pink and mauve hollyhocks which have followed us all the way from Zohab.

Another old friend who greeted us on the way was the hoopoo. What a vain fellow he is! He comes and alights on the path just in front of you, and deliberately fans out his pretty crest at you out of the purest conceit. His pride cost him dear in the old days, though, if the legend about him be true. The story is this. King Solomon, travelling in the desert at noon, was greatly afflicted by the heat, but sought in vain to find any shade. Presently a hoopoo flew by. He saw the great monarch in distress, and asked the reason of it. As soon as he learnt the cause, the little bird sped off and collected all his tribe, who, flocking together in the air above King Solomon's head, formed such a screen from the sun that he was able to take his noontide sleep in peace. When he awoke refreshed, he asked the hoopoos what reward they would choose, giving them leave to name whatever they most desired. After a long confabulation the hoopoos returned answer through their spokesman that they wished above all things to be given each a golden crown. At once a crest of pure gold grew from the head of each of the tribe. But their happiness was short-lived.

Though no one had paid attention to them before, they now became so valuable a prey that every man's hand was turned against them. In despair the survivors returned to King Solomon and begged that this blessing which had turned to a curse might be removed. "Well," said the King, "you've brought it on your own heads by your conceit, and you've no one but yourselves to blame for it. But I have not forgotten what you did for me in the desert, so I will save you from further persecution." At that moment the crown of gold on the head of each turned to a crown of feathers, and from that day the hoopoos have been once more free from molestation. But I am afraid they never really learnt that lesson!

But to come back again to our march through the fields and the oak-woods. As we zigzagged down the slope the Zâb came into sight. It was nearly as broad as where we had forded it forty miles lower down, and infinitely wilder, charging downwards through narrow gaps between the hills and foaming noisily over rocks and rapids. Instead of crossing it our path led us parallel to it up-stream, and for miles we climbed up and down over the spurs which run right down to its banks. It was bad country for the mules, but glorious for the traveller. Every valley was a delight with its wooded crags and splashing torrents, and glimpses far up and beyond of the great Kandîl range, a huge relentless wall of rock streaked with snow.

At the top of one rise, instead of dropping again,

as so often before, down into another gully, we came out on to a great broad plateau of corn-land with a sweeping view far away to the north, where a massive snow-mountain half shrouded in cloud formed a noble background. The fields were full of men reaping, and groups of chocolate-coloured Kurdish tents were dotted about everywhere. Here we camped, just outside a village. Each village has, at this time of year, its counterpart in the form of an encampment near by where the inhabitants pass the summer. The tents are very characteristic in shape. They are very long and low, and consist of a brown woollen material which ends in a long fringe near the ground, and, in the absence of any sort of ridge-pole, is poked up by the row of uprights inside into a series of shapeless knobs, each of which, in the best tents, is adorned with a large woollen tassel. The villages themselves are nothing but a collection of very wretched mud huts interspersed with mountainous heaps of dried dung-fuel, with very occasionally more ambitious buildings adorned with arches and loggias, which seem to point to the memory of better things in times long past.

To-day we marched on here through a string of such villages, out of every one of which huge white sheep-dogs rushed furiously at us as we passed, and were duly repulsed with whips and stones. They are ferocious-looking animals, but in reality very tame compared with the brutes you find in most parts of Asia Minor or the Balkans.

Mohammad Amîn Agha, a big local chief, came to call this afternoon. Every one else being away from camp either on survey work or erecting pillars, only B. and myself were there to represent the majesty of the British Empire when the *agha* and his suite arrived. I think B. rather bewilders these dignified old gentlemen who come to visit us, his methods with them being somewhat of the slap-him-on-the-back, "how are you, old buck?" variety, but they appreciate genuine cordiality, and the visit was quite a social success. Amîn Agha is a polite old Kurd with fairly polished and very agreeable manners. He sat for the sake of his dignity, but not at all at ease, on a mess-chair, while his followers squatted all round him. His little son of ten years old came too, and was vastly interested in a sparklet-syphon, till some one let it off in his direction and nearly frightened him out of his wits. The "grown-ups" we amused by showing them our rifles—a source of never-ending interest to any Kurd,—my own Mauser with its hair-trigger adjustment creating a great sensation, as they had never seen one of the kind before. We then produced an old '*Illustration*' with pictures of King George and Queen Mary; the King was much applauded for his fine robes, but the *agha* appeared a little shocked by the Queen wearing evening dress. Then it was their turn to show us their guns and daggers, some of the latter of fine workmanship. At this moment the mess *khitmagar* appeared in his ordinary green uniform and turban, and our guest (in whose country green is worn only by *seyyids*), jumping to the conclusion that our humble menial was a descendant of the Prophet, rose, seized his hand and kissed it fervently. The *khitmagar*, for fear of the complications which might otherwise ensue, was hastily prompted in Hindustani not to reveal that he was but an ordinary mortal, and submitted with a very sheepish look to the ordeal. We could not help feeling rather guilty, however innocent of the intention to deceive, especially as the *agha* had shown us a particular mark of courtesy when we first entered his district by riding out to meet us with 100 of his men, who acted as a sort of guard of honour, lining up on each side of the path as we passed along it.

Pasova, July 27.—To-day's march was entirely in the plain, a great change from the break-neck mountain tracks which have been our only roads since we left Kasr-i-Sherin. Our mules are amazingly sure-footed, and we have had only two casualties during all this time. The Persians have been less fortunate; only the other day four of their animals fell down a *kud*, and the "Victorious Leader," like Jill, went tumbling after, but happily sustained nothing worse than a sprained thumb. The poor Persian doctor fell too, but without hurt to his arm, I am glad to say.

This morning we crossed the Zâb for the last

time, almost at its head-waters. There was a village at the ford surrounded by the usual circle of haystacks, which provide fodder for the animals during the four winter months when the whole country is under deep snow, and a dwelling for the storks during the summer. Several village Rachels came down with great earthenware pots on their heads to draw water as we passed, and with them a small boy who looked delightfully comical in his very short shirt and the tall cap of the Mamâsh tribe, which is half-way between a pierrot's cap and a bishop's mitre. A little farther on we met the *agha* of the Mamâsh looking very fine with his men-of-war round him, though the effect was a little spoilt by the presence of an ordinary black umbrella held over his head by a mounted servant.

Ushnu, July 29.—By crossing an insignificant-looking watershed on our way here yesterday, we left the basin of the Tigris behind us for good and all and passed into the basin which drains into the lake of Urmia. From the top of the ridge we could just discern a straight line on the northern horizon which marked the lake. In the more immediate foreground we had a fine view of the valley of the Gadyr in which Ushnu lies. It is in reality a large plain, and, spread out as it was before our feet, it reminded one for all the world of a landscape by some old Dutch painter. The whole expanse was studded with little villages, each nestling in a grove of apricots, with an avenue of poplars leading out into the flat and treeless plain. Through the middle ran the Gadyr itself, lost among a maze of canals which branched off to each village in turn and fed the rich cornfields which spread from slope to slope.

Ushnu itself lies tucked away in the farthest corner of the plain, right under the mountains, half of it in the plain and half on the hillside. It is the largest town we have seen since Kasr-i-Sherin, and boasts a real Persian governor in uniform, and is garrisoned by a battalion of Russian cossacks, a token that we are now within the Russian "zone." The big man of the place, however, is the white-haired old chieftain of the Zerzaw, a sedentary Kurdish tribe who inhabit Ushnu and the plain. They are distinguishable from the other Mukri tribes by wearing, instead of a pointed cap, a most singular head-dress composed of a twisted rope of stuff wound round and round in an ever-narrowing spiral, so that it forms a covering to their heads which I can only compare to the sugar ornaments on a Christmas cake. I wonder if any country on earth can produce such a variety of sartorial absurdities as Kurdistan?

Later.—We have just come in from paying calls on the Governor and the old chieftain, Mansur-ul-Mamâlik (the Helper-of-States). The former received us in a favourite Persian way—that is, in a small, gaily-coloured tent pitched by the bank of a stream,—all Persians have a particular love for the sound of running water. After the formal two cups of tea, supplemented by the local dishes of vanilla ice and apricots, we went on to call on Mansur-ul-Mamâlik. His youngest son came to the outer door to meet us, a very merry-eyed youngster of twelve or fourteen, with such remarkably English features that it was hard to get rid of the idea that he was an English school-boy dressed up for a fancy-dress ball, for he wore a suit of dark green, with enormous trousers and a rolled cummerbund hiding the whole of his body. It was a bit of a shock to learn, as we did soon after, that he had a wife and was a confirmed gambler, losing his £20 or so at a sitting.

Musa Khan, as his name was, led us across a garden and up a flight of steps into his father's reception-room. It was a very long room, with mud walls containing a series of alcoves full of cushions for "company" to sit on, and hung with gauze curtains of pink and green. The spaces between the alcoves were occupied with oleographs, chiefly pictures of crowned heads, and specimens of Persian caligraphy gaudily framed. A foot or two below the ceiling ran a ledge all round the room on which was disposed the most wonderful array of china cups and dishes, lamps and bottles—enough to stock a china shop. On the wall facing me alone I counted more than sixty bottles; there were wine-bottles, beer-bottles, mustard-pots, spice-jars, and a dozen other varieties, all of them merely ornamental, I imagine, as the old fellow is a strict teetotaller. We ourselves were placed at the end of the room on a row of family chests, while the *agha* and his two sons, the elder a youth of twenty-three, squatted at our feet. They are a truly aristocratic family in birth, looks, and manners. The old man told us how only a few years before his written family history had been stolen and lost; it went back, he said, 250 years, and included the adventures of one of his ancestors who accompanied Nâdir Shah on his campaigns. There is nothing improbable in this, for the Kurds are an exceptionally pure-bred race, the tribal system, as among the Arabs, having the effect of preventing almost completely marriage with outsiders. Both the sons were strikingly handsome, with the large eyes and finely-moulded arched eyebrows that one knows so well in old Persian miniatures. Their father is a famous *raconteur*, and told us endless yarns, thumping his fist on his knee in ecstasy at the good points. Many of his stories were old Persian fables, one which he told with particular zest being the following :—

"A leopard strolling in the rice-fields met a cat. 'Hullo,' said he, 'what kind of a beast are you? You've got whiskers like mine, and a tail like mine, and stripes like mine, but you're so remarkably small; what in the world is the matter with you?' The cat was much hurt by the leopard's remarks, but he answered politely, 'Well, you see, I was like you once upon a time, but I fell in with humans, and now I live with them and have grown quite small and weak, as you see me.' 'Oh, and what sort of creatures are these humans, then?' asked the leopard. 'I'd like to meet one and show him what one of us can be like—will you introduce me?' 'All right,' said the cat, and off he went and fetched a man who was working in the rice. When the man came the leopard said to him, 'Look at me; I'm the same breed as your cat here, but I'm strong and I'm going to show you my strength; will you fight me?' The man was afraid to refuse, so he said, 'Yes, I'll fight you, but I left my strength at home this morning, and you must let me go and fetch it.' The leopard agreed and the man went away; but after he had gone a little distance he came back and said, 'That's all very well, but how do I know that you won't run away while I'm gone, and so give me all my trouble for nothing?' The leopard swore he would wait till the man returned, but the man said he could not trust him. 'Let me tie you up in my rice-bag so as to make sure,' he said, and after a little persuasion the leopard allowed himself to be tied up in the rice-bag. Then the man went home for his gun. 'Ah,' said the cat, who knew what would happen, 'you see it's just as I told you. You, too, have fallen in with humans, and you've become as small and weak as me.'"

I don't quite know where the moral lies, or if there is one, but the story as told in simple, vivid Persian by the old man was altogether delightful. After the story-telling the *agha* showed us his collection of Persian manuscripts. They were chiefly the verses of Hâfiz or Saadi, some of them bearing dates as far back as 1040 A.H.—that is, more than 300 years ago. The writing was exquisite, page following page with never a stroke or a curve in which one could detect the slightest imperfection, and written with the same meticulous care on the hundredth page as on the first.

At sunset (for we are in the month of Ramazan) dinner was brought in. The dishes were arranged on four huge trays which were set in a line down the middle of the room. We were invited to share the meal, and I have never tasted such kabâbs and delicious junkets since I was in Constantinople. For drink, a bowl of sour milk with lumps of snow floating in it went the round of the table, each person in turn using a large wooden ladle which was passed round with the bowl,—not strictly hygienic, I daresay, but one forgets that amid such surroundings. The different members of the household strolled in one by one and took their places at the common board, and so we left them—the courtly old aristocrat, surrounded by his sons and dependants, a striking picture of dignity and simplicity combined.

Ushnu, Aug. 2.—The whole Commission was entertained by the Russian officers of the garrison to-day to an *al fresco* luncheon-party. We rode through the town, along the bazaar which crosses the river in the form of a covered bridge like the

Ponte Vecchio at Florence, and out to an orchard where long tables had been improvised beneath awnings hung across from tree to tree. Our principal host was a very genial grey-haired captain, who spoke only Russian, but made up in action what he could not express in words by making frequent tours of the table and clinking glasses with each of us in turn. The lunch lasted for two and a half hours, and the fluid refreshment fulfilled the best Russian traditions of hospitality. Out of one and the same Persian tea-glass I drank (my neighbour gave me no choice) *vodka*, beer, red and white wines, *kvas*, and Russian benedictine, —and all this at 96° in the shade!

During the meal the soldiers provided a continuous entertainment which added great zest to the occasion. There was an infantry company, commanded by the jovial captain, and two separate detachments of Cossacks, one from the Caucasus and one from the Black Sea provinces. The infantrymen were clumsy, good-natured-looking peasants in the usual loose blouses and top-boots, but the Cossacks wore a very smart uniform consisting of a grey Astrakan *kalpak*, a rust-coloured coat with very long, full skirts, belted tightly at the waist and cut down to a point in front so as to show a sort of black "parson's waistcoat" underneath, breeches and riding-boots. The coat has a row of little pockets arranged in a slant across each breast, with a cartridge in each, and the men carry a long sheath-knife attached to the buckle of their belts in front—an awkward-looking piece of equipment. The officer's uniform is almost precisely the same, but he carries a sabre and has a row of silver dummies instead of real cartridges on his breast.

The men arranged themselves in circles, each man facing inwards, and sang. It was unaccompanied part-singing and beautiful to listen to. They tackled grand opera quite cheerfully, but they were best in their own Russian folk-songs. After a while a trio of soldiers appeared with guitars and played, sitting on the grass, while the Cossacks danced. It was the typical Russian dancing, much slapping of the thighs and boots, and the dancers bobbing down till they sat on the heels, and then shooting out one leg straight at front. Sometimes two of them danced together, back to back, every now and then clicking the soles of their boots against each other's in time to the music; and finally, two of their long knives were planted point-downwards in the ground and the Cossacks performed a sword-dance, each man holding his own knife with the blade between his teeth as he danced. The dancing was evidently as good fun for them as it was for us.

When the luncheon was over, we adjourned to a meadow to see an exhibition of trick-riding. Every single Cossack is an accomplished trick-rider, and some of their feats were really astonishing. One of the tricks consists of two men standing upright on their horses' backs at the gallop, while a third stands on *their* shoulders and holds the regiment's standard in his hand. Another favourite performance is for one man to gallop along with a led horse, vault off on to the ground, and then up again right over the back of his own horse on to that of the other horse on the farther side. To see these tricks, which would have surprised one at the Royal Military Tournament, being performed off-hand in a rough meadow, made one realise that the Cossack's reputation for horsemanship is founded on very substantial fact.

CHAPTER XV.

THE LAST STAGE.

WE left Ushnu on July 29, and on August 4, before we had reached Urmia, news of the outbreak of war reached us. It was the Russian Commissioner who first received it by means of a telegram from the commander of the Russian troops at Urmia. To say that it fell among us like a thunderbolt would be to use a mild simile, for—like most other Englishmen who found themselves at that moment in the remoter corners of the earth—we had heard not a whisper of warning beforehand. For the first few days all the details that reached us came either through the Russian official reports or else emanated from the "Agence Ottomane" from across the frontier. The news that flowed from these two sources was of the nature of an "alternating current,"—the Russian accounts being decidedly on the optimistic side, while the famous "Agence," needless to say, gave us the Berlin version of events undilute. It came about, for instance, that the latter having on one day disposed of the whole British Fleet at the bottom of the North Sea, we were consoled on the morrow by a rival rumour which resuscitated all our ships and replaced them on the bed of the ocean by eleven German Dreadnoughts! Our relief was great when, a few days later, we began to get our own regular supply of news in the shape of daily "Reuters" forwarded on by our Consul at Tabriz, who received them, only two or three days late, by way of India and Teheran.

Keen as everybody naturally was to return to his own country at such a time, there was nothing to prevent our work on the frontier continuing so long as none of the nations we represented were actually at war with one another. Turkey had indeed ordered a partial mobilisation at the outbreak of war, and we were prepared to hear of her participation at any moment; but though her two representatives, who were both officers on the active list, were obviously on tenterhooks, the Porte sent no orders to abandon the delimitation, so we merely forced the pace in the hopes of reaching the end before anything more happened.

The hundred miles or so of frontier which now remained to be settled followed the main watershed almost without a break the whole way to Ararat, and all the Commission had to do for the greater part of the distance was to put up pillars on each of the passes. The caravan was able, as a rule, to pursue a fairly level course along the edge of the plain, but the sub-commissions to whom it fell to erect the pillars, often had long and severe climbs to reach the crest. At least one of the passes was more than 10,000 feet above sea-level, and though we were well into August, and at about the latitude of Tunis, we had to cross snow-fields to reach the top, and bitter cold it was when we got there. The view on the farther side was a great contrast to the undulating plains of Azerbaijan from which we had ascended, the wild mountains of Turkish Kurdistan presenting a picture of nature in her grimest mood of upheaval. Even at these chilly heights, however, flowers grew in profusion. Great sheets of purple vetch spread between the snow-filled gullies, mingled here and there with some yellow alpine plant; but what particularly caught one's eye was a large thistle in colour of the deepest ultramarine—head, stalk, leaves, and all were of the same deep blue, so that the plant looked as if it had been dipped bodily into a dyer's vat.

On one of the highest of these passes, named Keleshîn, we came once again upon the handiwork of some frontier commission of an earlier age. At the very top of the pass, exactly on the line of our new frontier, stood a pillar — a monolith with an inscription in cuneiform characters. The existence of the pillar was already known, as it had been visited as much as eighty years ago by the archæologist Schultz, who was sent to Kurdistan and the neighbouring countries by the French Institute to examine all the monuments of antiquity and "arrow-headed" inscriptions which he might find. Indeed it is quite possible that this very inscription was the direct cause of that poor scientist's death, for it was in this neighbourhood that he was attacked and shot dead by the guards who had been provided for him by the Khan of Julamerk — the cause of whose gross act of treachery has always remained unknown.

This same pass of Keleshîn was in all probability the highroad from Nineveh to Ecbatana, and no doubt in the days of the kings of Assyria a considerable stream of traffic flowed along the track which is nowadays trodden by few except the local nomads. Tobias, it may be remembered, undertook this journey on his way to Ecbatana to court his fair cousin after his adventure with the fish on the banks of the Tigris. (As for the fish, by the way, Tobias's fears of being swallowed were less ill-founded than might appear, for the monsters that are caught near Mosul—the old site of Nineveh—are sometimes of such dimensions that when slung across a donkey's back their head and tail touch the ground on either side.)

It was at this stage that our Commission sustained a grievous loss in the person of the cook. The poor fellow, together with his compatriot the butler, had from the start protested against being obliged to ride, the Goanese having, it appears, a peculiar inaptitude for this method of locomotion — so much so, in fact, that on the first day out from Mohammerah they both tramped their weary twenty miles on foot, arriving in camp very woe-begone objects in their dusty black clothes. Next day the doctor—who, being mess secretary, was responsible for the quality of our dinner — was obdurate, and mounted them willy-nilly on muleback. Their fates thereafter proved as divergent as those of Pharaoh's cook and butler of yore; for whereas the butler finished up an expert and accomplished horseman, the cook had his worst forebodings realised when he toppled off the back of a pack-mule into a ditch, and in doing so fractured his collar-bone.

In spite of our anxiety to finish the frontier with the greatest speed, it was impossible to avoid a delay of some days at Urmia in order to refit and to rest the men and beasts; so we broke off work when we reached the district of Tergaver, and marched to Urmia across the plain. This plain of Urmia is perhaps the most fertile part of the Shah's dominions, and but for the Kurdish raiders would be the most prosperous. It is a large expanse of cultivated land fifty miles long by eighteen broad, producing great quantities of corn and tobacco, and lies between the mountains of Kurdistan and the lake which bears the same name as the town. We had no time during our short stay at Urmia to visit the lake; but its salt waters present such an unusual phenomenon that I cannot resist quoting the description given of them by the German traveller, Dr Moritz Wagner, who visited their shores in 1843.

"In the summer months, when the great lake is commonly as quiet as a pond in an English park, a deposit of mud results from the evaporation of the water. The prevalent colour of the water is blackish blue in the centre, and at a distance it appears azure, whereas close at hand it looks green and almost black, and so dense that fatty bodies, such as pigs, do not sink in it." He goes on to say that a chemical analysis of the water shows it to contain an immense number of ingredients resulting from the decomposition of water-plants, the mass of which he describes as being so great as to stop the breakers at some distance from shore. The learned doctor is then led to speculate as follows on the medicinal properties of the water: "If Lake Urmia were in the centre of Europe, our physicians would probably send thither thousands of their patients who could derive no benefit from the whole pharmacopœia, and who knows if a plunge in its waters might not renovate them. I, at all events, can affirm from personal experience that ten baths in the German Ocean do not afford so much stimulus to the skin or so much exhilaration to the nerves as the water of this lake, which holds so much more salt and iodine in solution than even the Dead Sea. You come out of its waters as red as a crab, and, moreover, greatly invigorated and refreshed. The Urmia baths would have this advantage over the North Sea —that its waves are not in the least dangerous, even in storms. Stout men who stretch themselves full length on its surface float without making any effort." One might almost infer, from his previous reference to the pigs, that the cautious doctor, before entrusting his own person to the waves, wisely tested their buoyant properties by means of the familiar *experimentum in corpore vilo!*

The town of Urmia is of great antiquity, and was of some fame even in the time of the Romans. Its chief claim to renown is, however, that it was there, or near there, according to the latest authorities, that Zoroaster himself was born. It is said that this founder of one of the oldest religions extant first taught along the shores of the lake—a curious parallel to those times, 700 years later, when the Author of Christianity likewise chose the side of a lake for the scene of His divine teaching. The last descendants of the Zoroastrians—or "fire-worshippers," as you sometimes see them named—linger on in Persia under the title of *Gabrs*, living chiefly in Yezd and Kermân, though large numbers of them flourish, of course, in India under the name of Parsis. Apart, though, from the legitimate adherents to the doctrines of the great prophet, it is an interesting reflection to one who has lately visited his birthplace that here was born that personality which, as metamorphosed by Nietzsche, gave rise to that conception of the "superman," which (if one is to believe many writers of the moment) first planted the seeds whose harvest is the present European War.

The town itself, set in the centre of endless gardens and tobacco-fields intersected by rows of tall French poplars, is of a considerable size, having extended appreciably since the Russians began to police the country and brought a hitherto unknown security of life and property to its inhabitants. There is a mile or so of fine vaulted bazaar running through the centre of the town, rich with the scents of leather and spices, and very grateful when one passes into its cool, voluptuous atmosphere from the glare and the heat outside.

The inhabitants of the town and the plain are, like those of the rest of Azerbaijan, of a race closely allied to the Ottoman Turks, and speak a language (Turki) which is practically an archaic form of modern Turkish, retaining the old original words which their ancestors brought with them from Central Asia, but which in "Stamboul" Turkish have been so often superseded by words borrowed from Arabic or Persian. One might perhaps roughly compare the difference to that between the English of the educated classes to-day and the dialects of those counties where the old Saxon words have remained most in use. The Azerbaijanis are thus entirely distinct by race and tongue from the Persians, although their country forms part of the empire. In the neighbourhood of Urmia there is also a large sprinkling of Nestorians, descendants (as far as their ancestry can be traced) of the ancient Assyrians. They are really mountain-folk living in the wild country across the border in Turkey, where their religious head—Mar Shimûm, as he is called—has his abode. Though, like the Copts in Egypt, they have to a great extent adopted the language of their more powerful neighbours, their own original Syriac is still in use among them, and is their usual medium for writing. Their relations with the Kurds are hardly more fortunate for themselves than are those of their fellow-Christians, the Armenians, and it is questionable which of the two has suffered most in the present war; but I shall come to the story of their persecution later.

The presence of the Nestorians seems to have been originally responsible for the arrival of American missionaries at Urmia in 1831, where they have maintained their mission—I think without a break—ever since. Wagner visited them in their early days, and has much to say about their kindness and hospitality, though he concludes with the peculiarly undiplomatic remark that "Mr Perkins [the head of the Mission] seems to be superior in character and intellect to his two colleagues," which invidious comparison he evidently considers amply atoned for when he adds that the latter "are, however, eminent for piety and virtue"! Without embarking on any such conscientious analysis as the German traveller, I can at least heartily endorse his appreciation of their spirit of hospitality, which added greatly to the pleasure of our visit to Urmia. But it is to qualities far beyond mere hospitality that the deepest tribute must be paid by any one knowing the story of these missionaries in the winter of 1914—namely, bravery and self-sacrifice equal to anything that the war has brought to light. First, however, I must give some account of what happened at Urmia six weeks after we left.

Long before Turkey declared war, the Turks incited the Kurdish tribes on their side of the frontier to descend into Persia and attack the towns of Azerbaijan which were garrisoned by Russian troops. At that time there were at Urmia several members of the Archbishop of Canterbury's Mission to the Assyrian Christians (a Mission which came into being rather more than twenty years ago in response to an appeal by the Nestorians—or "Syrians" as they are sometimes, rather ambiguously, called—to the English Primate, asking him to send out priests of the Church of England to help in the regeneration of their Church, which had fallen far away from the doctrines inherited from its foundation in 450 A.D.) The missionaries from the Turkish side of the frontier having had to leave their stations, they were all gathered in their Urmia quarters when the Kurdish invasion took place, of which the following vivid account was published by Mr MacGillivray, the head of the Mission, in their quarterly review :—

"On October 1 the real trouble began. On that day a large force of Kurds came down to Tergavar, drove out the small force of Cossacks, and started plundering and burning the Syrian villages. Most of the Syrians escaped and fled to the city, but a few were left behind and killed. The next ten days were a reign of terror. Every day swarms of Kurds poured down upon the plain. Every night some village or another was attacked, pillaged, and burnt, and each night they drew nearer to the city. Every night we heard a continuous fusillade, and saw the burning villages around us, while every day refugees poured into the city. All the Missions, as well as private houses of Christians, were full of them. We ourselves had about 350 on our premises. First we filled all our spare rooms, and then we let the others camp in the yard until every square foot was occupied.

"It soon became evident that this was no ordinary Kurdish raid, but the attack of an organised army several thousand strong. It was, in fact, a deliberate expedition planned and organised by Turks (egged on, no doubt, by Germans), whose object was to drive out the

Russians and take Urmia. There were Turkish officers among the Kurds, and they had German ammunition. Moreover, they had an agreement with the Moslems of the city, who, when the Kurds entered, were to rise and join with them in the plunder and massacre of the Christian quarter. It was also clear that the small force of Cossacks was quite insufficient to deal with the enemy. The Russian Consul assured us that reinforcements were coming, but day after day passed and no reinforcements appeared. Then the Russians raised a very useful additional force by serving out rifles and ammunition to the Syrians. We even got half a dozen rifles and a box of cartridges ourselves, ready to guard our house, hoping that the Archbishop's embargo on priests bearing arms did not apply to self-defence against a horde of brigands.

"The climax was reached on Sunday, October 11. All day we saw large bands of Kurds coming down the mountain-slopes. Besides the Cossacks the Russians had a few guns in the city, and with these they shelled the enemy as they approached. From our roof we could see the shells dropping and exploding among them. This checked the attack for some time; but in the evening after dark the enemy came on again, and that night made a very determined attack on Charbash—a village not more than half a mile from the city wall. Firing began at ten o'clock in the evening, and continued all night; but the Russians and

Syrians together made a good defence, and in the morning the enemy retired, leaving many dead.

"This was the last attack. It is generally believed that the attack on the city itself was fixed for Monday night, and behind the enemy's firing line were large numbers of women and children with baskets and sacks, all ready to carry off the plunder. But they never got their plunder. On Monday we again saw large bands of Kurds on the mountain-slopes, but they were going the other way. All day long we saw them creeping up from the plain and hastily retiring. This was explained in the evening, when news came that large Russian reinforcements—infantry, artillery, and machine-guns—had arrived, and were encamped a few miles off. The Kurds did not care to await their coming."

Soon after this abortive attack on Urmia the English missionaries were ordered to return to England, and when on January 2, 1915, the Russian troops evacuated the town, and the Turkish troops and Kurdish tribesmen poured into the place, the American missionaries alone were left. The Kurdish tribesmen having the Christian population now at their mercy, committed every possible sort of atrocity, burning, as is reported on good evidence, over 100 villages, and slaughtering as many of the wretched Nestorians as fell into their clutches. 'The Near East,' in a letter written from Urmia in July 1915, gives the following description of what took place in the invaded country :—

"A massacre took place in the village of Gulpashan, where fifty men, after being tied arm to arm by the soldiers, aided by the native Moslems, were taken out to the graveyards and were there butchered like animals. In another case some forty-six men were taken out from the French Mission. After the persons in authority had given assurance that they were to be transported to Turkey, they were tied together and were shot at a place about two miles from the city. They were killed in cold blood, without any pretence at any kind of trial. One of them who escaped told the following story: 'We were tied together arm to arm, and made to kneel and await our doom. They fired at each one of us. Soon after the firing I began to feel around my body to see whether I was shot. I finally came to the conclusion I was not hit, but I fell with the rest, making pretence that I was dead. After we had all fallen they came near, and stabbed every one who seemed to be still breathing, or showing the slightest sign of life in him. As they came near me they could not find any sign of life, for I made believe I was dead long ago, and giving me a kick, they left us. After they had departed a long way from us I got up and made my way to the American Mission.'"

The writer of this letter places the number

of Christian refugees who took asylum with the missionaries (who fortunately have very large premises, with spacious courtyards and gardens) at nearly 18,000, of whom "nearly 2000 were lodged for weeks and months in the church without room to lie down." One woman who had been there for three months, and had nothing but a narrow wooden seat to sit upon, was asked if she was not tired of remaining always in the same spot. "Oh no," she answered, "this is a good place. See, I have a place for my head." She had a pillar behind her which she could rest her head against! One of the missionaries had gone out at the beginning, when the massacre was in full swing, and interviewed the leader of the Kurds, thereby saving the lives of 1000 villagers, whom he brought safely to the mission. Another who ventured outside the mission walls was seized and, in spite of his nationality, severely beaten. But the real heroism of this little colony—there were, I think, eight or nine men, most of them with their wives—came to the fore when the inevitable epidemic broke out among this packed crowd of refugees. The death-rate in their courtyard reached 40 a day, and it was with great difficulty that the missionaries could even obtain leave from the Turks to take the corpses out for burial. Practically all of them, men and women, caught typhus from the sick whom they tended, and more than half their number died.

It was not until May 20th that the Turkish

troops and their Kurdish auxiliaries withdrew, and the Christian population were able to return to their ruined homesteads. Only a small remnant of their race can have survived the Kurdish massacres, the epidemic in the American Mission, and the enormous mortality among those who, when the Russian forces withdrew from Urmia, attempted in the very depth of winter to reach the Russian frontier on foot. In Urmia itself hardly a single Christian escaped except those whom the gallant Americans saved.

A few miles to the south of Urmia, a solitary mountain, named Seer, rises out of the plain, and provides a cool retreat in summer when the heat in the town below becomes excessive. The Russian garrison, consisting of a regiment of infantry (the 5th Caucasian Rifles), a battery of guns, and a detachment of Cossacks, were camped on its slopes when we arrived. One afternoon we drove out along a road many inches deep in white dust to call on the officers of the mess. We intended to pay an ordinary afternoon call, but we had yet to learn the full meaning of Russian hospitality—when we left Seer it was after midnight, with the regimental band playing us off, and an escort of Cossacks, with flaring torches, to see us safely home. From six o'clock till eleven we sat at table, commencing with a sort of high tea, and drifting imperceptibly into a many-course dinner. The fact that none of us spoke a word of Russian, nor any of our hosts a word of anything else, heightened rather than lessened the merriness of the proceedings, for, not being able to talk, we sang. Heaven knows how many times over we sang our own and each other's National Anthems, with the Marseillaise occasionally sandwiched in; and when at last we had satiated our patriotic ardour, our hosts began to sing queer, plaintive songs of the Caucasus, half Russian, half Turkish, and rousing toasting songs, with a special verse for each guest. Presently, in answer to a bugle-call outside, the Colonel and his officers rose, and, excusing themselves for a few minutes' absence, left the room. They had gone out to join the men at evening prayers, and after a few minutes the notes of a hymn came floating in from the hundreds of voices outside. After they had returned, a number of the soldiers themselves trooped into the long, bare mess-room (for the regiment was mobilised, and everything packed up except the actual chairs and tables). The Caucasians brought in with them a strange-looking brass trophy fixed to the end of a pole, reminding one rather of those which the Roman legionaries carried, but that it had a couple of dyed horse-tails and dozens of little bells attached to it. Together with a guitar and cymbals, it provides the music for their songs on the march, and the noise which this primitive orchestra gave forth when shut in by four walls was deafening.

The soldiers stood in a ring round the orchestra and sang with all their souls, their coarse, rather brutish faces lit up and almost transfigured by their enthusiasm for the music. After they had sung for some time, the Cossacks began to dance. At first they were the same dances that we had seen at Ushnu, with the same slapping of boots, clicking of heels, and clashing of knives. In spite of their long riding-coats and top-boots these little horsemen were amazingly nimble, and soon the pace became furious. Their officer, a thin, grey man with shaven head and prodigious moustaches, was sitting with us, and I could see his eye gleaming as he grew more and more excited at the dancing; suddenly he could contain himself no longer—with one bound he jumped from his seat at the table to the centre of the floor, and the next moment he was dancing away more madly than any of them.

Some of the Cossack dances are really a rude sort of play in which the dancers act simple parts, such as that of the coy maiden being wooed by her ardent lover. In one, more elaborate than the rest, a Cossack sat alone in the centre of the floor whittling away at a stick with his long sheath-knife, and crooning to himself. A dozen of his companions came and formed a ring round him, crouching on the ground with their heads bent low and their hands shading their eyes, and chanting a dirge of misery. Suddenly the man in the middle sang two lines, in solo, apparently to the effect that—

> "What's the good of melancholy?
> Life was made for joy and dancing,"

for with a whoop they all leapt to their feet and started the wildest dance imaginable, singing, clapping and slapping their thighs in a very ecstasy of motion.

All this dancing and singing was very strange to Western eyes. It is so palpably inspired by a national spirit quite different to our own, that to me, at least, it seemed to typify somehow that great, mysterious Russia, which, however many books are written about it, remains to most of us so unknown — perhaps so unknowable. Yet in that bare, whitewashed barrack dimly lit by a few oil-lamps slung from the roof, listening to the barbaric music, and watching the uncouth but fascinating movements of the dancers, I felt a momentary flash of insight into the real Slav spirit such as no other surroundings could conceivably have produced.

After a bare week at Urmia, we were back again at work on the frontier, skirting along the foot of the mountains, as before, and only stopping to set up pillars on the passes. On August 21 we witnessed the interesting spectacle of an eclipse of the sun. The path of the eclipse, passing through the Crimea and Bagdad towards

India, would, we knew, nearly intersect our line of march, and it was a little uncertain whether we should find ourselves within the belt of the shadow or not. The interest was therefore intense, as we sat with darkened telescopes and bits of smoked glass watching the black rim of the shadow creep over the sun, and hoping for the glorious possibilities of coronas and polar rays. The orb gradually diminished till nothing but the thinnest of thin shreds was left, the light faded to twilight, it grew very cold, and strange luminous ripples began to race over the ground; then the shadow slowed down, remained stationary for a minute or two and, to our grievous disappointment, began to recede. We had missed a total eclipse by a few miles.

A few days later the Commission reached the small town of Dilmân. On the way there we passed by the stronghold of a somewhat notorious Kurdish chief, whose name was Ismaïl Agha, but who was usually known by the curiously sounding nickname of "Simko." It was, thanks to a freak of nature, one of the most peculiar places I remember having seen. Through the middle of a level plain ran a deep, square-cut ravine like a gigantic trench. It was perhaps 300 yards broad and 100 feet deep, but remained quite invisible till one came almost up to its brink. In the centre of this ravine a great mass of rock stood up like an island, its summit a little below the level of the plain. It formed a perfect natural citadel, and some chieftain had seized on it many years ago, and built a castle on the top, so cunningly contrived that it was next to impossible to see where the sheer rock finished and the walls of masonry commenced. Round the foot of the rock a village clustered, as villages used to gather round the old feudal castles in the days of the Barons. "Simko," however, proved himself a very degenerate baron, for under the influence of civilisation he had abandoned the romantic home of his ancestors, and was busy building a banal residence on the edge of the plain, and—worst horror of all—was installing a telephone!

From Dilmân we headed again into the mountains, and climbed to an altitude of 7000 feet.

Barely 100 miles now lay between us and our goal; already a cloud-wrapt peak, dimly seen from a summit some days before, had been identified as the great Ararat itself. And here let me make my bitter confession. In spite of my promises to the patient reader, in spite of the title of my book itself—I never set foot on Ararat! My only sight of it was from the window of a railway carriage, and all my hopes and ambitions of climbing its snowy heights were destined to ruin. The reason you shall learn before I close this book, but first let the Venerable among Mountains have his due. I must needs borrow for the purpose, for I have no phrases of my own to describe the majesty of which I had but a momentary glimpse. The following words reveal, however, the effect which it produced on an English traveller of forty years ago :—

"Towering above all and soaring up into a firmament so clear as at once to convey to the mind, or rather to the imagination, an idea of infinite space, a rugged and solitary pyramid of eternal snow dwarfs by comparison every neighbouring or visible headland. It is Mount Ararat.

"The Turks call it Agridagh, or Mountain of the Ark; the Persians Koo-i-noo, or Noah's Mountain; and the Armenians Massees, or Mother of the World.

"Moses of Khorene pronounces it the middle of the world; and both Raumer and Hoff maintain that it is the central point of the great terrestrial line drawn from the Cape of Good Hope to Behring Straits.

"Having seen the mountain of the Deluge from several points of view, some of which looked up its actual sides, I am of opinion that nowhere else on the face of the earth is there a mountain whose effects on the mind of the beholder can be compared to it.

"Many other mountains in the world are much higher; but although Ararat is only 17,210 feet above the level of the sea, it soars without a rival or a neighbour—a solitary pyramid or cone, 10,876 feet above the flat plain in which it stands. It is this circumstance which endows it with such overpowering majesty."[1]

[1] From Creagh's 'Armenians, Koords, and Turks.'

The same writer relates these interesting traditions connected with Ararat:—

"As the superstitious Kurds and Armenians believe that the Ark, still painted green and resting on the extreme summit, is guarded by Jins — devils or evil spirits — nothing in the world will persuade any of them to ascend its sides beyond a certain height.

"A monk very long ago indeed attempted to climb up, in order that his piety might be whetted by the contemplation of a piece of the Ark which he proposed to bring away with him; but, although employing several days in the journey, he was at length obliged to desist, for in the evenings, when falling asleep upon the mountain-side, some supernatural agency carried him back to the point from which he had started in the morning.

"To reward him for his pains, however, an angel brought him down a piece of the Ark, and informed him at the same time that since the landing of Noah no human being had ever, or ever would be, allowed to visit the place of his disembarkation."

The obstacle which prevented me from reaching our goal was of a far more concrete nature than that which the good old monk encountered—namely, a Kurdish bullet. On August 31 five of us were attacked by local tribesmen while we were shooting partridges within a mile or two of camp; and though my companions escaped injury—most miraculously, seeing that our assailants fired several scores of shots, some at a range of little over twenty yards—I had the misfortune to be hit. Therewith ended my connection with the Commission. It was my great good fortune to be in the skilful hands of Captain Pierpoint, and together we returned direct to England—as direct, that is to say, as is possible in times of European war, for our journey lasted no less than nine weeks. An improvised litter carried by the Indian *sowars* was my first conveyance, which we exchanged on reaching the level plain for a rough Russian ambulance, to be replaced in turn, as soon as we reached Khoi and a so-called road, by the most modern type of motor ambulance sent from the frontier to meet us by the Russian Commandant. At Julfa, the frontier town and railhead of the Caucasus railway, the Belgian Director of Customs offered us the hospitality of his house, while the Russian officials, to whom I have cause to be everlastingly grateful, provided a special coach for us for the two-days' journey through Tiflis to Batoum on the Black Sea. From there we took a Russian steamer to Constantinople, touching at all the Turkish ports on the southern shore of the Sea on the way, and being held up for a day at the mouth of the Bosphorus — thanks to the attentions of the *Goeben*, who, with her consort the *Breslau*, was diverting herself with target practice just outside the entrance to the straits. We reached Constantinople to find the Dardanelles already closed, though war was not yet declared between Turkey and the Allies; but by taking train to the Bulgarian port of Dedeagatch we were able to get on board a British steamer which crept down the Syrian coast to Egypt, "and so home," as Pepys would say, landing there early in November—just a year from the time we had started. Incidentally we had, in the course of this year, made the exact circuit of the Turkish dominions in Asia.

Three weeks after the attack on our party, the Commission, now dwindled down to very small numbers, reached the point on Mount Ararat where the Turco-Persian frontier now joins up with the frontiers between these two countries and the Russian Empire. When the two officers who remained (the others had gone back to India) started to return to England, they found every way closed to them except Archangel; and so, for the first time probably since the days of those early merchant adventurers, Englishmen followed the route from Persia to their native land *viâ* the Arctic Ocean.

AFTERWORD

By Susan Littledale

Dorothy never did get to see Mosul. Nor did GE. The injuries inflicted by the Kurdish bullet, and the Turkish entry into the war on the side of the Germans, combined to bring a premature end to his work in the Levant. His last letter home was dated 29 August 1914, two days before the attack, but fortunately the full story of the incident – and his colleagues' long journey home after reaching Ararat – can be found elsewhere.

The Commission arrived in the village of Ashnok in north-west Iran on the morning of 31 August 1914. Having pitched camp near the village, four members of the British delegation set out to shoot partridges accompanied by ten unarmed sowars acting as beaters. They proceeded up a nearby ravine and the party separated as the ravine split in two. Some time later, as they returned to the camp, eight to ten Kurdish tribesmen appeared 'on the slopes of the spur between the two ravines'[1] and opened fire on one of the groups. Miraculously unharmed, they hurried back to camp, where they found all the members of the other group safe and well. All, that is, except GE – the only Commission member without a shotgun.

A rescue party was sent out and found him lying wounded on the spur. They brought him in an hour later. A bullet had penetrated his thigh, severing the sciatic nerve and causing complete paralysis of the lower leg. In his preliminary report the Commission Medical Officer, Captain H. W. Pierpoint, who had accompanied the rescue party, noted that

> there is complete paralysis of all muscles below the knee [. . .] the wound of the thigh is exquisitely painful, the pain being referred entirely to the foot, and at times being so agonizing as to need morphia for its control.[2]

The details of the shooting are undisputed, although the reasons behind it are less clear. We know this from a Foreign Office file in the National Archives containing sworn depositions to a 'court of enquiry' conducted by Captain Wilson four days after the incident. All members of the shooting party described a deliberate and unprovoked attack by up to ten Kurdish tribesmen, some shooting at them from a distance of fifty yards. In his statement Deputy Commissioner Colonel Ryder reported that: 'the fire was deliberately aimed, several shots passing quite close to us. I could see them clearly and there could be no doubt that they could also see us clearly'.[3]

In his own statement on the incident GE recalled that he had been crossing the spur when he saw 'a group of eight armed men on the further side of the *nullah* [ravine] shouting and running about'.[4] Seconds later he felt himself hit in the leg. Unable to walk, he fell to the ground. After a while several Kurds came up to him and questioned him in a dialect difficult for him to understand. All he could make out was that they had clearly thought he was a Russian. Surprised by his insistence that he was not, and now 'evidently in doubt what to do',[5] his attackers asked a few more questions before they departed leaving him still lying on the ground.

The area where the shooting took place was controlled by the Russians. In 1907 the Russians and the British, taking advantage of political instability in Persia, had formalised three zones of influence under the Anglo-Russian Convention of 1907. This north-west corner of Persia was in the Russian zone and the local tribal chief, Ismail Agha (Simko), was responsible to the Persian government for this part of the frontier with Turkey. Simko, the head of the Kurdish Shikak tribe, who was 'in receipt of a salary of 4000 Tomans per mensem as Warden of the Marches'[6] for the Kotur, Dilman and Somai districts, clearly had some questions to answer.

Six of the tribesmen were brought before the enquiry and questioned by Captain Wilson. While two of them admitted firing shots at GE none of them repeated the claim that they thought he was a Russian. The one who had shot him said he had thought GE was a Turkish soldier and had shot him because 'we hate the Turks and fear them'.[7] A second, who had shot once and missed, said he had done so because 'we came to the conclusion it must be an enemy'.[8] Simko, who had rounded up the culprits and brought them to the enquiry at the Commissioner's request, was present when they were publicly and separately interrogated.

Captain Wilson observed in his report to the Foreign Office that 'the excuse [. . .] advanced by them that they thought they [the shooting party] were Turks',[9] seemed plainly to be an afterthought. It was admitted in evidence that they had heard firing for an hour without taking any action and 'the party was dressed distinctively and could not have been mistaken for Turks'.[10] He was in no doubt that

> the outrage was committed by the tribesmen of the Muhammadi Shikak clan with the object of embroiling Ismail Agha (who has frequently looted them) with his friends the Russians, to whom he owes his present position and influence, Russian friendship having taken the substantial form of a present of 1,000 rifles with ammunition.[11]

Whether this was an accurate interpretation of the facts must remain a matter for conjecture. What was clearly established, however, was that the attack was completely unprovoked and that it was carried out by a Persian tribe on Persian soil. It was the Persian government, Wilson argued, that should take responsibility for the administration of justice and he made it known that he recommended 'that two of the accused, who admitted firing at Mr Hubbard, should be hung and that the rest should be severely beaten and imprisoned for a period'.[12]

* * *

GE's journey home, outlined in the final pages of his book, was followed by a long and painful period of recuperation. In the next fifteen months he was to undergo three major operations and seven weeks of treatment in a special nursing home in an effort to save his leg. While he wrote *From the Gulf to Ararat* in the periods of enforced convalescence between operations, a battle to get him a proper level of compensation was being fought in the background.

Initiated by Foreign Secretary Sir Edward Grey, and later taken up by Lord Curzon, former Viceroy of India and Persia expert, it began with a claim for full compensation from the Persian government and a demand that the two Kurds be executed. However, the political situation in Persia soon became too delicate to push for the ultimate sanction. Sir Edward was advised by our man in Tehran that the Persian Minister for the Interior felt that 'agitation in Azerjaiban makes it most unsuitable to carry out executions at the moment'.[13] Whether the executions ever took place appears open to question. In 1917 the Persians paid out £2,000 in compensation, more than half of which went on medical costs. In 1919 the doctors gave up the struggle to save GE's leg and it was amputated. As Lord Curzon, now Acting Foreign Secretary in the post-war coalition government, fought the Treasury's reluctance to pay him an allowance 'analogous to the wound pension granted to military officers'[14] GE embarked on rebuilding his career.

As for the other expedition members, most of them dispersed into the theatre of war as soon as the expedition ended. The last protocols and maps of the frontier had been signed near Bayazid at the foot of Mount Ararat on 27 October 1914. By this time the only remaining members of the British delegation were the Commissioner, Captain Wilson, and his deputy, Colonel Ryder. They, too, would soon resume military duties, but first there was the not inconsiderable task of getting the fruits of their labours back to England.

The Commissioner later recalled his last hours in camp. Colonel Ryder had gone ahead with the precious documents while he stayed on to pack up the tents and baggage. A few hours after Ryder had left, the Turkish delegation suddenly raised objections to Wilson's departure. Conscious that a declaration of war between Turkey and the Allies was imminent, and that he was in danger of being taken prisoner, Wilson had to think on his feet.

'I instantly not only agreed to stop but, saying I was in no hurry, invited the Turkish officials to lunch with me next day.'[15] He returned to his own camp and, telling the local servants not to prepare food for him that evening as he was dining with the Turks, he walked out into the darkness and made for the frontier, where he met up with Colonel Ryder. When the Turks arrived for lunch the next day they found a sumptuous meal but no Captain Wilson, who was by this time well out of reach. Almost a year passed by before he discovered that, upon finding him gone, 'the Turkish officer laughed – "he has cheated us – but not of our dinner" and they sat down to enjoy themselves; after which they took charge of my tents and animals'.[16]

Wilson and Ryder, who had driven from Maku in a *droshky*[17] and crossed the Oxus two hours before Turkey declared war on Russia, spent the next ten days travelling by train to Archangel (Arkhangel'sk) in north-west Russia. Here they just managed to catch the last ship to leave that winter before the seas froze up. In the company of crates of Russian and Siberian eggs and piles of timber, they passed through the Norwegian fjords to Bergen and finally disembarked with the cargo in Hull. Once back in London their first stop was the Foreign Office, where they handed over 'our copy of the *Carte Identique* and of our protocols, and our accounts complete to date'.[18]

* * *

There seems to be little doubt that GE's work on the Commission, and his employment by the Foreign Office whilst recuperating in London, were compelling factors in his favour when he sought advancement into a diplomatic post after the war. In February 1920 he was formally transferred to the Foreign Office and six months later was appointed Second Secretary at the British Legation in Peking. Promoted to First Secretary in 1922, he resigned in June 1924 to take up a post in the Peking office of the Hong Kong and Shanghai Banking Corporation.

GE was not a banker and was not employed as a member of the foreign staff, who would typically be recruited as school leavers and trained as generalists in the London office before being sent 'East'. In fact, his name doesn't appear on staff lists for this period as he was employed on a special contract as the bank's political advisor. As a specialist his role was to negotiate with the Chinese government on the issue of loans and other matters.[19]

He stayed with the bank until 1933. During his 13 years in China, in addition to mastering the language, he developed an expertise in Far Eastern affairs, which was eventually to lead him to the post of Far Eastern Research Secretary at the Royal Institute of International Affairs (Chatham House) in London. From there, now a recognised authority in his field, he was seconded to the Political Intelligence Department of the Foreign Office during World War II.

His work at the Foreign Office ended in June 1943 and he returned to Chatham House until his retirement in April 1945. Never one to retire in the full sense of the word, his final years were taken up researching the history of the village of Wye in Kent[20] where he lived until his death on 16 May 1951.

Notes

1 National Archives, T164/4/13. Hubbard, G.E., Secretary to Turco-Persian Commission; Compensation paid by Persian Government and ex-gratia payment from Indian British Governments for leg injury when fired upon by Kurd tribesmen. Statement by Lieutenant-Colonel C. H. D. Ryder, 4 September 1914.

2 National Archives, T164/4/13. Preliminary Medical Report by Commission Medical Officer H. W. Pierpoint.

3 National Archives, T164/4/13. Statement by Lieutenant-Colonel C. H. D. Ryder, 4 September 1914.

4 National Archives, T164/4/13. Statement by G. E. Hubbard, 5 September 1914.

5 Ibid.

6 National Archives, T164/4/13, Captain A. T. Wilson report to Sir Edward Grey, 7 October 1914.

7 National Archives, T164/4/13. Translation of evidence given to enquiry before Captain Wilson at Ashnok, 4 September 1914.

8 Ibid.

9 National Archives, T164/4/13., Captain A. T. Wilson report to Sir Edward Grey, 6 September 1914.

10 Ibid.

11 Ibid.

12 National Archives, T164/4/13. Captain A. T. Wilson report to Sir Edward Grey, 7 October 1914.

13 National Archives, T164/4/13. Telegram from Sir W. Townley (Tehran) to Sir Edward Grey, 24 November 1914.

14 National Archives, T164/4/13. Earl Curzon to the Secretary to the Treasury, 15 August 1919.

15 Arnold Talbot Wilson (1942), *S. W. Persia: Letters and Diary of a Young Political Officer 1907–1914*, Readers Union, London, p. 300.

16 Ibid., p. 301.

17 An open carriage.

18 Wilson, *S. W. Persia*, p. 304.

19 HSBC Group Archives, HSBC Holdings plc, London.

20 G.E. Hubbard (1951), *The Old Book of Wye*, Pilgrim Press Derby.

INDEX.

THE END.